ATOMIC WEIGHTS BASED ON CARBON-12

								O 18
			IIIB 13	IVB 14	VB 15	VIB 16	VIIB 17	HELIUM 2 **He** 4.0026
			BORON 5 **B** 10.81	CARBON 6 **C** 12.011	NITROGEN 7 **N** 14.0067	OXYGEN 8 **O** 15.9994	FLUORINE 9 **F** 18.9984	NEON 10 **Ne** 20.179
	IB 11	IIB 12	ALUMINIUM 13 **Al** 26.9815	SILICON 14 **Si** 28.0855	PHOSPHORUS 15 **P** 30.9738	SULPHUR 16 **S** 32.06	CHLORINE 17 **Cl** 35.453	ARGON 18 **Ar** 39.948
10	COPPER 29 **Cu** 63.546	ZINC 30 **Zn** 65.39	GALLIUM 31 **Ga** 69.72	GERMANIUM 32 **Ge** 72.59	ARSENIC 33 **As** 74.9216	SELENIUM 34 **Se** 78.96	BROMINE 35 **Br** 79.904	KRYPTON 36 **Kr** 83.80
NICKEL 28 **Ni** 58.69	SILVER 47 **Ag** 107.868	CADMIUM 48 **Cd** 112.41	INDIUM 49 **In** 114.82	TIN 50 **Sn** · 118.71	ANTIMONY 51 **Sb** 121.75	TELLURIUM 52 **Te** 127.60	IODINE 53 **I** 126.905	XENON 54 **Xe** 131.29
PALLADIUM 46 **Pd** 106.42	GOLD 79 **Au** 196.967	MERCURY 80 **Hg** 200.59	THALLIUM 81 **Tl** 204.383	LEAD 82 **Pb** 207.2	BISMUTH 83 **Bi** 208.980	POLONIUM 84 **Po** 209	ASTATINE 85 **At** 210	RADON 86 **Rn** 222
PLATINUM 78 **Pt** 195.08								

GADOLINIUM 64 **Gd** 157.25	TERBIUM 65 **Tb** 158.925	DYSPROSIUM 66 **Dy** 162.50	HOLMIUM 67 **Ho** 164.930	ERBIUM 68 **Er** 167.26	THULIUM 69 **Tm** 168.934	YTTERBIUM 70 **Yb** 173.04	LUTETIUM 71 **Lu** 174.967
CURIUM 96 **Cm** 247	BERKELIUM 97 **Bk** 247	CALIFORNIUM 98 **Cf** 251	EINSTEINIUM 99 **Es** 252	FERMIUM 100 **Fm** 257	MENDELEVIUM 101 **Md** 258	NOBELIUM 102 **No** 259	LAWRENCIUM 103 **Lr** 260

ELLIS HORWOOD SERIES IN INORGANIC CHEMISTRY
Series Editor: J. BURGESS, Department of Chemistry,
University of Leicester

IONS IN SOLUTION: Basic Principles
of Chemical Interactions

JOHN BURGESS, Department of Chemistry, University of
Leicester

This outline of inorganic solution chemistry treats an
important chemical area which is only rather cursorily
covered in most undergraduate textbooks. It also serves
as an introduction to the author's more comprehensive
classic *Metal Ions in Solution* (Ellis Horwood Limited,
1978). A feature of this new text is its lucid treatment, both
of simple cations and anions and of a selection of complexes, to provide a balanced coverage of inorganic
species.

The first half of the book covers fundamental topics such
as the extent and nature of solvation of ionic species in
solution. The second half deals with such aspects of
thermodynamic and kinetic behaviour as acidity, polymerisation, solvent exchange, complex formation, redox
properties, and electron transfer reactions. Much of the
book is devoted to aqueous solutions, but non-aqueous
media feature predominantly in appropriate sections.

The book is written for first or second year undergraduates with a major interest in chemistry, and would also
be useful for more advanced undergraduates or research
workers in related sciences. It is paired with two cassettes
produced by and available from the Educational Techniques Subject Group of the Royal Society of Chemistry.

Readership: Inorganic chemists; first and second year undergraduates and research workers in related sciences.

IONS IN SOLUTION:
Basic Principles of
Chemical Interactions

J. BURGESS, MA, PhD
Department of Chemistry
University of Leicester

ELLIS HORWOOD LIMITED
Publishers · Chichester

Halsted Press: a division of
JOHN WILEY & SONS
New York · Chichester · Brisbane · Toronto

First published in 1988 by
ELLIS HORWOOD LIMITED
Market Cross House, Cooper Street,
Chichester, West Sussex, PO19 1EB, England
The publisher's colophon is reproduced from James Gillison's drawing of the ancient Market Cross, Chichester.

Distributors:

Australia and New Zealand:
JACARANDA WILEY LIMITED
GPO Box 859, Brisbane, Queensland 4001, Australia

Canada:
JOHN WILEY & SONS CANADA LIMITED
22 Worcester Road, Rexdale, Ontario, Canada

Europe and Africa:
JOHN WILEY & SONS LIMITED
Baffins Lane, Chichester, West Sussex, England

North and South America and the rest of the world:
Halsted Press: a division of
JOHN WILEY & SONS
605 Third Avenue, New York, NY 10158, USA

South-East Asia
JOHN WILEY & SONS (SEA) PTE LIMITED
37 Jalan Pemimpin # 05–04
Block B, Union Industrial Building, Singapore 2057

Indian Subcontinent
WILEY EASTERN LIMITED
4835/24 Ansari Road
Daryaganj, New Delhi 110002, India

British Library Cataloguing in Publication Data
Burgess, John, *1936–*
Ions in solution.
1. Ions. Solution
I. Title
541.3'72

Library of Congress Card No. 88–662

ISBN 0–7458–0172–2 (Ellis Horwood Limited)
ISBN 0–470–21059–1 (Halsted Press)

Phototypeset in Times by Ellis Horwood Limited
Printed in Great Britain by Butler & Tanner, Frome, Somerset

Table of contents

Author's preface

The general area of inorganic solution chemistry is treated rather cursorily in many undergraduate textbooks. A number of readers of the author's *Metal Ions in Solution* have over the years suggested that an undergraduate level version would be a useful teaching aid. The initial response was to plan a pair of cassettes with accompanying workbooks, to be produced by the Educational Techniques Subject Group of the Royal Society of Chemistry. The first of these appeared in 1984, the second is in preparation. Meanwhile it has been decided to make this material available in conventional textbook format. The first draft of the manuscript for this book consisted simply of the scripts for the cassettes. Subsequent versions have involved a little rearrangement and considerable rewriting, but the scope and level of the text remain very similar to those of the cassettes. The solution chemistry of simple anions and of a selection of complexes has been added to that of metal cations, to give a more balanced coverage of inorganic species. Topics dealt with include the extent and nature of solvation, and some spectrscopic, thermodynamic, and kinetic characteristics of inorganic ions in solution. Much of the book is devoted to aqueous solutions, but several sections reflect greater knowledge of certain aspects in non-aqueous media. Some basic knowledge of inorganic and physical chemistry is assumed, such as that acquired in the first year of an Honours Chemistry course. A Glossary of some fundamental terms has been included in order to help readers with limited background knowledge. Lists of Further Reading direct the reader to fuller accounts of certain areas and also provide him/her with an introit into more detailed or advanced treatments. The material of the book should in turn provide a basis from which specialised final-year courses can be developed, either in pure chemistry or in one of the ever-increasing number of joint or combined degree courses.

The list of grateful thanks to some of the many people who have helped in various ways to guide this book from inception to publication should start with Bob Gillard, Professor at University College Cardiff. It was his invitation to survey the current state of this area of chemistry at the RSC Autumn Meeting held at University College Cardiff in 1980 which led to a series of lectures around the country. This in turn, specifically the lecture at Hull, led to the suggestion of the preparation of a cassette and workbook. I am grateful to Dick Moyse for initiating and encouraging this idea, and to Peter Groves for his subsequent enthusiasm and patience in coaxing the first cassette into being. I am very grateful both to Peter Groves and to Ellis Horwood my publishers for their cooperation in arranging for parallel production of this textbook, and to Ellis himself and his experienced and dedicated staff for all their efforts and care in its preparation. Finally it gives me pleasure to acknowledge my debt to my colleagues at Leicester and to several generations of undergraduates and research students there, whose support, interest, and enquiring minds have done so much to improve my knowledge of chemistry, in solution and indeed in general.

Leicester
December 1987

John Burgess

List of symbols and abbreviations

aq	water; aquated; aqueous medium
C	heat capacity (specific heat)
	C_P isobaric (constant pressure) heat capacity
	$\Delta C_P^{\ddagger}$ heat capacity of activation
CFAE	crystal field activation energy
CFSE	crystal field stabilisation energy
d^n	d-electron configuration of a transition metal ion
D	dissociative (mechanism)
Dq	crystal field splitting parameter
E	energy; redox potential
	$E^{\ominus}$ standard redox potential
e_g	see Glossary, under 'Crystal Field'
EXAFS	extended X-ray absorption fine structure
$\mathscr{F}$	Faraday
$g(r)$	radial distribution function
G	Gibbs free energy
	$\Delta G^{\ominus}$ standard Gibbs free energy change
	$\Delta G^{\ddagger}$ Gibbs free energy of activation
h	Planck's constant
H	enthalpy
	ΔH_{hydr} enthalpy of hydration
	$\Delta H^{\ddagger}$ enthalpy of activation
	$\Delta H(\text{M–L})$ metal–ligand bond dissociation energy
HSAB	Hard and Soft Acids and Bases
I	interchange mechanism
I_a	associative interchange
I_d	dissociative interchange
k	rate constant
	k_0 rate constant at atmospheric pressure
	k_1 first-order rate constant (dissociative or solvolysis path)

k_2 second-order rate constant (associative path)

k_b rate constant for reverse (back) reaction in an equilibrium

k_{ex} rate constant for solvent (ligand) exchange

k_f rate constant for complex formation *or* rate constant for forward reaction in an equilibrium

k_P rate constant at high pressure

K equilibrium constant (see also pK below)

K_n stability constant for the addition of the nth ligand in complex formation

K_{os} outer-sphere association constant

n number of ligands in a complex *or* number of electrons

P pressure

pK negative logarithm to base 10 of equilibrium (stability) constant K (analogous to pH)

R alkyl

R gas constant

S entropy

$\bar{S}^{\ominus}$ standard partial molar entropy

$\Delta S^{\ddagger}$ entropy of activation

SCS   sterically controlled substitution (mechanism)

S_N nucleophilic substitution

S_N1 unimolecular, i.e. dissociative

$S_N1(lim)$ limiting dissociative mechanism

S_N2 bimolecular, i.e. associative

T temperature

t_{2g} see Glossary, under 'Crystal Field'

V volume

$\bar{V}^{\ominus}$ standard partial molar volume

$\Delta V^{\ddagger}$ volume of activation

$\Delta V_i^{\ddagger}$ volume of activation for ligand interchange

$\Delta V^{\ominus}$ standard volume change for a reaction

$\Delta V_{os}^{\ominus}$ standard volume change for outer-sphere pre-association equilibrium

w_{ij} work term (in bringing reactants, particularly in redox reactions, together)

z charge on an ion

β_n stability constant for the addition of n ligands to a metal ion

ν frequency

ABBREVIATIONS FOR LIGANDS AND SOLVENTS

Lower-case letters are used in this book for ligand abbreviations, upper-case for solvent abbreviations. Such compounds as dimethyl sulphoxide can act in either capacity: their typographical appearance varies accordingly.

Ligands which are anions of weak acids are taken as L^{n-}: the free organic molecule is then LH_n.

aa⁻	amino-acid anion (see ala⁻, asp⁻, gly⁻)	
acac⁻	acetylacetonate(pentane-2,4-dionate)	$[MeCOCHCOMe]^-$
ala⁻	alaninate	$[MeCH(NH_2)CO_2]^-$
asp⁻	aspartate	$[HO_2CCH_2CH(NH_2)CO_2]^-$
bipy	2,2'-bipyridyl	
3CNacac⁻	3-cyanoacetylacetonate	$[MeCOC(CN)COMe]^-$
cp⁻	cyclopentadienyl anion	
DMA	NN-dimethylacetamide	$MeCONMe_2$
DMF/dmf	NN-dimethylformamide	$HCONMe_2$
DMSO/dmso	dimethyl sulphoxide	Me_2SO
edta⁴⁻	ethylenediaminetetraacetate (ethane-1,2-di-aminetetraacetate)	
en	ethylenediamine (ethane-1,2-diamine)	$H_2NCH_2CH_2NH_2$
gly⁻	glycinate	$[H_2NCH_2CO_2]^-$
HMPA/hmpa	hexamethylphosphor(tri)amide	$OP(ONMe_2)_3$
L,LL,LLL,..	general symbols for mono-, bi-, tri-dentate ligand	
4,7-Me₂phen	4,7-dimethyl-1,10-phenanthroline	
Nu, nucl	nucleophile	
ox²⁻	oxalate	$[O_2CCO_2]^{2-}$
oxinate	8-hydroxyquinolinate	
pada	pyridine-2-azo-4'-dimethylaniline	
PC	propylene carbonate (4-methyl-1,3-dioxalan-2-one)	
phen	1,10-phenanthroline	
py	pyridine	
terpy	2,2',6',2''-terpyridyl	
THF	tetrahydrofuran	
TMP	trimethyl phosphate	$OP(OMe)_3$
TMTU	tetramethylthiourea	$SC(NMe_2)_2$
TMU	tetramethylurea	$OC(NMe_2)_2$
tu	thiourea	$SC(NH_2)_2$
X⁻	halide	

1

Introduction

1.1 DISSOLUTION OF SALTS

One of the best ways to appreciate the importance of ion–solvent interactions in electrolyte solutions is through the cycle shown as the top half of Fig. 1.1. This relates

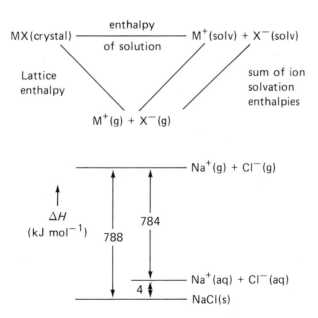

Fig. 1.1 — Interrelation of solution, solvation, and lattice enthalpies.

the enthalpy of solution to ion solvation enthalpies and to lattice enthalpy. Invariably the enthalpy of solution of a salt is the small difference between the large enthalpy needed to separate the ions from each other in the crystal lattice and the enthalpy gained when these ions are introduced into the solvent. The lower half of Fig. 1.1 shows that for the specific case of sodium chloride, the enthalpy of solution is only about 0.5% of the lattice or ion solvation enthalpies. Fig. 1.1 should give some idea of the strength of ion–solvent, particularly ion–water, interactions. In the following pages various aspects of the chemistry of solvated ions — their natures, properties, and reactions — are introduced.

Solvent molecules can become attracted to ions with varying degrees of firmness, depending of course on the characteristics of both the ion and the solvent. The introduction of ions into a solvent can also have a marked effect on its properties. This is particularly true when, as in the case of water, the solvent has a pronounced structure of its own. Indeed the commonest and most important solvent, water, is one of the most interesting in this respect, since in aqueous solutions of salts interactions between ions and solvent molecules profoundly affect interactions between the solvent molecules themselves. Many sections will deal mainly with aqueous solutions, as these have received most attention, owing both to the importance of water and to its interest as a solvent. However, the study of non-aqueous solvents has developed greatly in some areas, and such solutions will be discussed where appropriate.

Much of this book, especially the earlier chapters, will be concerned with the nature and properties of metal ions, though much of the discussion is equally relevant to anions and to complex ions. The main reason for this imbalance is simply that some of the more fundamental aspects of the chemistry of ions in solution are better documented and understood for solvated metal ions than for other solute species. However, an aqua-metal ion is in reality only a special case of a complex, with water acting as ligand (see section 1.4).

1.2 METAL IONS AROUND THE PERIODIC TABLE

Before embarking on the various aspects of the chemistry of solvated ions, it may prove helpful to summarise the distribution of aqua-metal ions in relation to the Periodic Table. In Fig. 1.2, elements giving one or more well-established species of this type are shown with a tinted background. There are also several much less well-established aqua-cations, some of which are listed in Table 1.1. Many of these have been postulated in connection with the measurement of physical properties such as redox potentials or stability constants. Simple representations such as Au^+, Bi^{3+}, or Zr^{4+} are really shorthand for species which are more complicated and indeed are very difficult to characterise properly. Difficulties arise from strong tendencies to hydrolyse and polymerise, to form complexes, to disproportionate, or to oxidise or reduce solvating water.

Salts dissolve to a greater or lesser extent in a range of polar solvents, both hydroxylic solvents such as the alcohols and dipolar aprotic solvents such as acetonitrile or dimethyl sulphoxide. In all these cases the cations will be solvated, as in aqueous solution. Methanol, ethanol, and acetone are rather less effective in solvating metal ions than water, but dimethyl sulphoxide or pyridine solvate some cations considerably more effectively than water. In general metal ions which give hydrated cations in water can usually give analogous solvated cations in polar organic solvents. Simple inorganic anions, on the other hand, such as halides or oxoanions,

```
H                                                                      He
Li  Be                                          B   C   N   O   F   Ne
Na  Mg                                          Al  Si  P   S   Cl  Ar
K   Ca  Sc      Ti  V   Cr  Mn  Fe  Co  Ni  Cu  Zn  Ga  Ge  As  Se  Br  Kr
Rb  Sr  Y       Zr  Nb  Mo  Tc  Ru  Rh  Pd  Ag  Cd  In  Sn  Sb  Te  I   Xe
Cs  Ba  La  Lnt Hf  Ta  W   Re  Os  Ir  Pt  Au  Hg  Tl  Pb  Bi  Po  At  Rn
Fr  Ra  Ac  Acttt
```

† Lanthanide cations
†† Actinide cations

Fig. 1.2 — Distribution of the elements that give one or more well-established solvated cation species — such elements are indicated by a tinted background.

Table 1.1 — Dubious, difficult, and potential aquacations

Ill-characterised aquacations
   Cited in redox potential data: Au^+aq; $Au^{3+}aq$; $Sn^{4+}aq$; $Pb^{4+}aq$
   Cited in thermochemical tables: $Ga^{2+}aq$; $Bi^{3+}aq$; $Zr^{4+}aq$.

Aquacations that react with water
   Hydrolysis and polymerisation: $Bi^{3+}aq$; $Sn^{4+}aq$; $Zr^{4+}aq$
   Disproportionation: Cu^+aq; $Mn^{3+}aq$; $Ga^{2+}aq$
   Oxidise water: $Co^{3+}aq$; $Au^{3+}aq$; $Pb^{4+}aq$
   Reduce water: $Ln^{2+}aq$ (Ln = lanthanide (4f) element)

Possible future aquacations
   $Os^{2+}aq$; $Os^{3+}aq$; $W^{3+}aq$; $Tc^{3+}aq$; $Tc^{4+}aq$; $Re^{4+}aq$

tend to be weakly solvated in organic solvents. It is this feeble anion solvation which so often makes simple inorganic salts very sparingly soluble in organic solvents. To increase the chances of high solubility in organic media the choice of anion should fall on an essentially hydrophobic anion such as tetraphenylboronate.

1.3 NEW AQUA-METAL IONS

It should not be thought that all possible metal ions able to exist in aqueous solution have already been discovered and characterised. In the last few years aqua-ions of palladium(II), platinum(II), and molybdenum(III) have, rather belatedly, been properly established. There is no reason to believe that aqua-ions of, for example, rhenium(IV) and technetium(III) and (IV) may not soon be characterised, as suggested at the foot of Table 1.1. The types of approach which have proved successful are illustrated in Fig. 1.3, which shows the methods of preparation used for

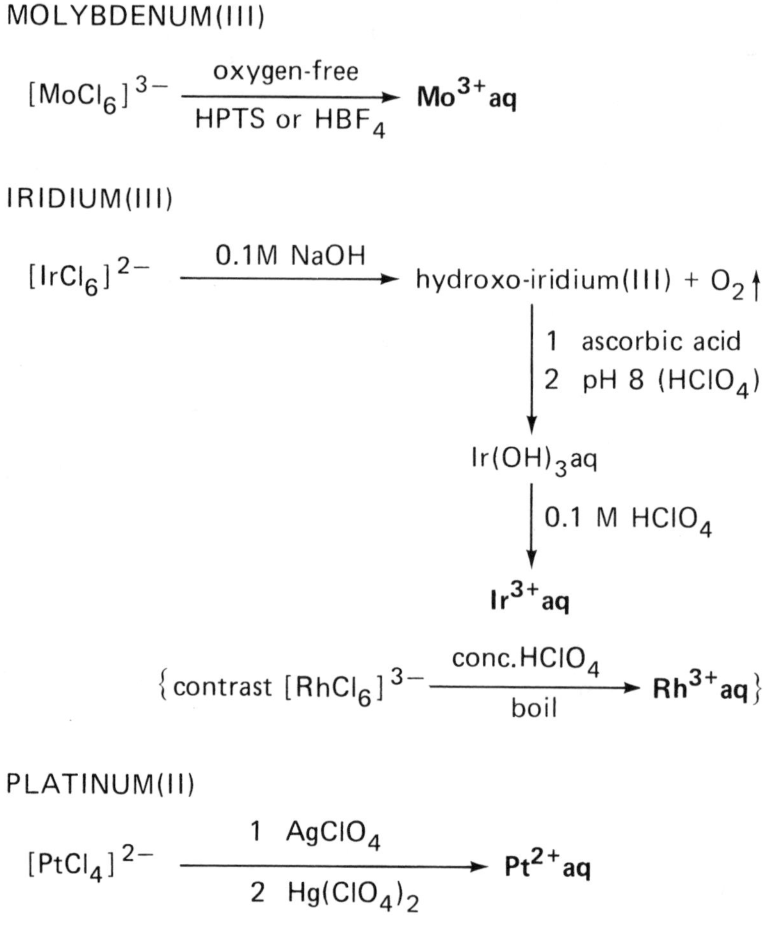

MOLYBDENUM(III)

$$[MoCl_6]^{3-} \xrightarrow[\text{HPTS or HBF}_4]{\text{oxygen-free}} Mo^{3+}aq$$

IRIDIUM(III)

$$[IrCl_6]^{2-} \xrightarrow{\text{0.1M NaOH}} \text{hydroxo-iridium(III)} + O_2 \uparrow$$

$$\downarrow \begin{array}{l} 1 \quad \text{ascorbic acid} \\ 2 \quad \text{pH 8 (HClO}_4) \end{array}$$

$$Ir(OH)_3 aq$$

$$\downarrow \text{0.1 M HClO}_4$$

$$Ir^{3+}aq$$

$$\{ \text{contrast } [RhCl_6]^{3-} \xrightarrow[\text{boil}]{\text{conc.HClO}_4} Rh^{3+}aq \}$$

PLATINUM(II)

$$[PtCl_4]^{2-} \xrightarrow[\text{2 \ Hg(ClO}_4)_2]{\text{1 \ AgClO}_4} Pt^{2+}aq$$

Fig. 1.3 — Preparative routes used for recently characterised new aqua-metal ions.

generating aqua-ions of molybdenum(III), platinum(II), and iridium(III).† The preparative methods shown in Fig. 1.3 illustrate several important points, especially in relation to complex formation. If it is necessary to add acid to control pH, the anion of the acid added must not form a complex with the potential aqua-cation. Chloride forms quite stable complexes with many metal ions, and thus hydrochloric

† Aqua-ions of iridium(IV) and iridium(V) are under investigation at the time of writing.

acid is to be avoided. From this point of view perchloric acid, p-toluenesulphonic acid, and trifluoromethane sulphonic acid are to be preferred, but are still not ideal. It has been demonstrated recently that there are significant interactions between p-toluenesulphonate and lanthanide cations, while the coordination of trifluoro-methylsulphonate to such metals as tin, iron, and palladium is well-established through X-ray structural studies. Indeed there are now extensive kinetic results on solvolysis and base hydrolysis of $[M(O_3SCF_3)(NH_3)_5]^{2+}$ cations, with M = e.g. Co, Rh, and Ru. In like vein, there is much evidence for complex formation involving perchlorate; titanium(IV) perchlorate, $Ti(ClO_4)_4$, is in fact an uncharged complex with four bidentate perchlorate ligands firmly bonded to the titanium. Tetraphenyl-boronate, $[BPh_4]^-$, is a non-coordinating anion, but has the disadvantage of forming rather a large number of salts which are practically insoluble in water. It is also not particularly stable in acidic solution. However, it is useful for avoiding complex formation and minimising ion-pairing in solutions containing hydrophobic cations in organic solvents, where its hydrophobic nature enhances solubility (see section 1.2). Even better is the $[B_{11}CH_{12}]^-$ anion, in 1986 awarded the title of 'least coordinating anion'. In organic solvents of low dielectric constant and low anion solvating power, perchlorate is quite liable to form complexes, as are ions such as PF_6^-, BF_4^-, or $CH_3SO_3^-$. These are all more reluctant to form complexes in aqueous media. The other requirement for the anion in the present context is that it should not undergo redox reactions with the solvated metal ion being sought. In this respect perchlorate (like nitrate) can be at a disadvantage, as it is reduced by several strongly reducing metal ions, such as V^{2+}, V^{3+}, Ti^{3+}, and Mo^{3+}. Nonetheless, despite these limited misgivings over perchlorate, it is in practice often the anion of choice when a relatively inert and non-complexing anion is needed. This choice stems from its ready availability and its frequent use by earlier workers in this field.

The use of mercury(II), often in the form of its perchlorate, is in a way complementary to these considerations, for its very strong affinity for chloride means that it has often been used to remove chloride ligands from a metal ion. In the absence of complexing anions perforce the ion generated becomes solvated to give the required aqua-cation. Similar remarks apply to silver(I), which removes chloride as insoluble silver chloride. The use of this approach is illustrated in Fig. 1.3 in relation to aquaplatinum(II), prepared from the readily available starting material $K_2[PtCl_4]$.

The other point of general importance in Fig. 1.3 is the need to consider possible redox complications. It was mentioned above that perchlorate and nitrate might have undesirable oxidising effects. It may be necesary to keep oxygen away from strongly reducing species. If the cation produced oxidises or reduces water, then in principle a stable aquacation cannot be obtained. However, some aquacations that are thermodynamically capable of oxidising or reducing water actually react very slowly and can in practice be isolated. Examples include Cr^{2+}aq and Ru^{2+}aq. Even Co^{3+}aq and Cu^+aq persist long enough for spectroscopic and kinetic studies to be feasible. Of course other solvates of such cations may well be stable to redox decomposition and readily isolable.

Potential applications of these principles are outlined below in suggested routes to the as yet uncharacterised aquarhenium(IV) and aquaplatinum(IV) cations. Potassium hexachlorohenate is a convenient starting material for the former:

$$K_2[ReCl_6] \xrightarrow{Hg(ClO_4)_2;\ HClO_4} Re^{4+}aq$$

Exclusion of air (oxygen) would be essential, in view of the strong tendency of rhenium(IV) to become oxidised to perrhenate (ReO_4^-; rhenium(VIII)). Aquaplatinum(IV) could perhaps be prepared from the relatively recently characterised aquaplatinum(II):

$$Pt^{2+}aq \xrightarrow{Cl_2} [Pt(OH_2)_5Cl]^{3+} \xrightarrow{Ag^+\ or\ Hg^{2+}} Pt^{4+}aq$$

The aqua-ions discussed so far are reasonably stable and inert. In recent years a number of strongly reducing and strongly oxidising aquacations have been generated by chemical methods (e.g. $Co^{3+}aq$; Cu^+aq; In^+aq) and by techniques such as pulse radiolysis (e.g. Zn^+aq; Cd^+aq; Ni^+aq; $Pb^{3+}aq$; $Th^{3+}aq$). Most of these aquacations have only a transient existence, and our knowledge of their properties is often restricted to their ultraviolet-visible absorption spectrum and their rate constant for decay.

Finally, in this discussion of the existence of solventocations in solution, it should be emphasised that we know much less about such species in non-aqueous media than in aqueous media. However, in a few areas, for example those of the determination of solvation numbers and rates of solvent exchange, there is actually more known about non-aqueous systems than about aqueous systems. Given a sufficiently reduction-resistant polar non-aqueous solvent it is even possible to generate solvated alkali metal anions. The macrocyclic polyether 15-crown-5 (Fig. 1.4) is a liquid and an effective ligand for alkali metal ions (see section 6.4);

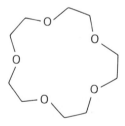

Fig. 1.4 — The cyclic polyether ligand 15-crown-5.

alkali metal anions can be stabilised by complexation in this distinctly unusual solvent. Simpler non-aqueous solvents such as liquid ammonia, liquid sulphur dioxide, and liquid hydrogen fluoride have such unpleasant and inconvenient properties that there is a considerable dearth of information on many aspects of these interesting and important inorganic non-aqueous media. There is still a need for a great deal of research into the nature and properties of solutions of electrolytes in these solvents, and indeed into electrolyte solutions in many more tractable but still neglected organic solvents.

1.4 ANIONS AND COMPLEXES IN SOLUTION

The situation in relation to anions is rather different, as mentioned at the end of section 1.1. In the first place there are very few monatomic anions in aqueous solution, essentially only F^-, Cl^-, Br^-, and I^-. There are a large number of important oxoanions, but evaluation and discussion of their solvation characteristics is complicated by the variety of stoichiometries and stereochemistries — for instance planar NO_3^-, pyramidal ClO_3^-, and the multitude of tetrahedral species such as ClO_4^-, MnO_4^-, and SO_4^{2-}. Halogenoanions include tetrahedral (e.g. BF_4^-) and octahedral (e.g. PF_6^-, $IrCl_6^{2-}$) species; octahedral $[Fe(CN)_6]^{3-}$, $[Fe(CN)_6]^{4-}$, and $[Cr(NCS)_6]^{3-}$ are derived from the linear pseudo-halides CN^- and NCS^-. It is to polyatomic ions of the MO_4^{n-} or MX_6^{n-} type that one has to turn in order to investigate effects of varying charge on properties of solvated anions in the absence of simple X^{n-}aq for $n > 1$. These two families of ions can also furnish a wide range of radii, and indeed of other properties as they are formed from a variety of elements, metals and non-metals, from most parts of the Periodic Table. All $[MO_4]^{n-}$ and $[MX_6]^{n-}$ (X = halide, cyanide, or thiocyanate) anions are, however, hydrophilic. If one needs a large hydrophobic anion, then one can turn to the tetraphenylboronate anion, $[BPh_4]^-$. This has the ions $[PPh_4]^+$ and $[AsPh_4]^+$, and, slightly less hydrophobic, the series of tetraalkylammonium ions $[NR_4]^+$, as complementary cations. These essentially organic ions are often useful in conferring solubility in organic solvents. Their large size, small charge, and therefore limited solvation in polar solvents proves valuable in establishing single ion thermodynamic parameters (section 4.3).

The various $[MX_6]^{n-}$ anions mentioned in the preceding paragraph are examples of complex ions. Such anionic species are complemented by a very large number of cationic, and a relatively small number of uncharged, complexes. All the elements with any metallic character form complexes in one or more of their oxidation states. The most electropositive elements, the alkali metals, have the lowest tendency to form complexes, especially in aqueous solution. The slightly less electropositive alkaline earth cations (Ca^{2+}, Sr^{2+}, Ba^{2+}) and lanthanide cations form complexes somewhat more readily. The transition metals and the earlier members of the actinide series form an enormous number of complexes. A wide variety of anionic and polar molecules can act as ligands. A very limited selection of examples is given in Table 1.2. The majority of complexes consist of a metal ion surrounded by a number of ligands. The most common coordination numbers (stereochemistries) are four (tetrahedral or square-planar), six (octahedral), and eight (square-antiprismatic or dodecahedral). However, coordination numbers range from two to twelve inclusive, with an isolated example of fourteen-coordination at uranium, and a variety of stereochemistries are known, especially for coodination numbers seven and eight. If a ligand has more than one lone pair of electrons available for donation then it may act as a bridge between two metal ions, to give binuclear or polynuclear complexes (Table 1.3).

There is a voluminous literature on complexes and their solution chemistry. In this book we shall only deal with aspects directly related to our main theme. One

Table 1.2 — Examples of ligands

Group IV	Group V	Group VI	Group VII
CN^-; CO	NH_3; NO_2^-; NCS^-	OH^-; ONO^-	F^-
	N_3^-; NCO^-; NR_2^-	CO_3^{2-}; SO_4^{2-}; NO_3^-	
		OR^-; RCO_2^-	
	PR_3; $P(OR)_3$	S^{2-}; SCN^-; SR^-	Cl^-
	AsR_3	$SeCN^-$; SeR^-	Br^-
$SnCl_3^-$	SbR_3		I^-

Table 1.3 — Bridging ligands and binuclear and polynuclear complexes

Ligands	Complexes
Monatomic	
Cl^-; Br^-; I^-	$Ag_2Br_3^-$; Ag_2I^+; Ag_3I^{2+}; $Ag_6I_8^{2-}$
Polyatomic	
OH^-	$\left[(en)_2Co \overset{OH}{\underset{OH}{\diagup\diagdown}} Co(en)_2 \right]^{4+}$
NH_2^- (with OH^-)	$\left[(H_3N)_4Co \overset{NH_2}{\underset{OH}{\diagup\diagdown}} Co(NH_3)_4 \right]^{4+}$
NCO^-	$[(H_3N)_5Cr-NCO-Cr(NH_3)_5]^{5+}$
CN^-	$[(H_2O)_5Cr-NC-Co(CN)_5]$
pyrazine	$\left[(OC)_5W-N\underset{\smile}{\frown}N-W(CO)_5 \right]$

general point that should be made is that an aquacation such as $[Cr(OH_2)_6]^{3+}$ is just a special case of a complex ion. There is no fundamental difference between $[Cr(OH_2)_6]^{3+}$ and such species as $[Cr(NH_3)_6]^{3+}$ or $[Cr(NCS)_6]^{3-}$, though the fact that ligand and solvent are identical for an aqua-complex does have important consequences. We shall deal at some length later with thermodynamic (Chapter 6) and kinetic (Chapters 9 to 12) aspects of complex formation and dissociation, i.e. the replacement of coordinated solvent molecules by other ligands and the reverse. The area of binuclear and polynuclear complexes is also relevant to the polymerisation of aquacations (Chapter 5) and certain classes of electron transfer reactions involving aquacations (Chapter 12).

Preparative methods for complexes generally fall within the province of the solution chemistry of ions. The majority of preparations of complexes are carried out in solution, often simply involving the reaction of an aqua-metal cation with the ligand concerned, itself often anionic. The hardest part is often not the generation of the required complex, but its isolation in the form of a pure solid. Fortunately there is always a wide choice of counterions available. Some random examples of simple preparations of complexes, showing the counterion of choice, are given in Table 1.4.

Table 1.4 — Preparations of complexes from aqueous solution

Metal ion	Ligand (in excess)	Counterion	Product
Ni^{2+}	ammonia	fluoroborate	$[Ni(NH_3)_6](BF_4)_2$
Ni^{2+}	ethane–1,2–diamine	thiosulphate	$[Ni(en)_3](S_2O_3)$
Co^{2+}	thiocyanate	mercury(II)	$Hg[Co(NCS)_4]$
Hg^{2+}	iodide	potassium	$K_2[HgI_4]$
Fe^{2+} Fe^{3+}	cyanide	potassium	$\begin{cases} K_4[Fe(CN)_6] \\ K_3[Fe(CN)_6] \end{cases}$

The choice of counterion in any particular case is often a matter of intuition or luck. The rule-of-thumb guideline that the counterion should be of equal and opposite charge and of similar size provides some help in selection. There is a sound theoretical basis for this rule-of-thumb, which derives from the balance between the lattice energy of a salt and the solvation energies of its constituent ions. An interesting application of this principle relates to the nickel(II)–cyanide system. In concentrated aqueous cyanide solution the equilibrium

$$[Ni(CN)_4]^{2-} + CN^- \rightleftharpoons [Ni(CN)_5]^{3-}$$

is established. By adding $[Cr(NH_3)_6]^{3+}$ or $[Cr(en)_3]^{3+}$, the $[Ni(CN)_5]^{3-}$ anion can be precipitated, providing a convenient route for the isolation of this anion.

Sometimes the aqua-ion M^{n+}aq is not conveniently available as starting material. Thus, for example, Co^{3+}aq survives only for a few minutes in aqueous solution as it rapidly oxidises water, while Cu^+aq disproportionates 'instantly'. For the very important case of cobalt(III), complexes can often be prepared simply by oxidising a solution containing Co^{2+}aq and ligand(s) with air or hydrogen peroxide. Care often has to be exercised to get the experimental conditions exactly right, or a complex slightly different from that required may be obtained. Some examples of cobalt(III) complex preparations are outlined in Table 1.5.

Table 1.5 — Preparations of cobalt(III) complexes

Cobalt(II) salt used	Ligands added	Oxidation conditions	Product
chloride	ammonium chloride aqueous ammonia	air or H_2O_2; charcoal catalyst	$[Co(NH_3)_6]Cl_3$
		air or H_2O_2	$[CoCl(NH_3)_5]Cl_2$
nitrate	ammonium carbonate aqueous ammonia	air	$[Co(CO_3)(NH_3)_5](NO_3)$
sulphate	ammonium carbonate	H_2O_2	$[Co(CO_3)(NH_3)_4]_2(SO_4)$
acetate	sodium nitrite aqueous ammonia	air	$[Co(NO_2)_2(NH_3)_4](NO_2)$
acetate	ethane-1,2-diamine oxalic acid; HCl	lead dioxide	$[Co(ox)(en)_2]Cl$
carbonate	potassium oxalate oxalic acid	lead dioxide	$K_3[Co(ox)_3]$
nitrate	sodium bicarbonate	H_2O_2	$Na_3[Co(CO_3)_3]$
chloride	potassium cyanide	boil in air	$K_3[Co(CN)_6]$

The Cr^{3+}aq cation presents different problems as a starting material for the preparation of chromium(III) complexes. This cation is readily available (though not from a bottle of hydrated chromium(III) chloride, which contains mainly $[CrCl_2(OH_2)_4]Cl)$, and is redox stable. But it is substitution inert, forming complexes extremely slowly, so it is often easier to prepare chromium(III) complexes by redox reactions. One route parallels that just discussed for cobalt, to oxidise a solution containing Cr^{2+}aq, itself easily prepared by reduction of Cr^{3+}, and the ligand(s). Another approach makes use of the powerful oxidising powers of chromium(VI). If chromate or dichromate are treated with an excess of an organic ligand, the product is often the chromium(III) complex of the ligand. A good example of this approach is provided by the reaction between chromium(VI) and oxalic acid, which gives $[Cr(ox)_3]^{3-}$ or $[Cr(ox)_2(OH_2)]^-$ depending on conditions.

An important group of elements where aqua-ions M^{n+}aq are rarely available as starting materials for the preparation of complexes comprise the noble metals, such as gold, platinum, iridium, and osmium. Treatment of ore concentrates with such aggressive reagents as aqua regia or molten bisulphates gives various chloro-, nitrato-, or sulphato-complexes. These species are then converted by ligand replacement or substitution reactions into the required complexes. This sequence is illustrated for platinum in Fig. 1.5. The metal is dissolved in aqua regia, to give the

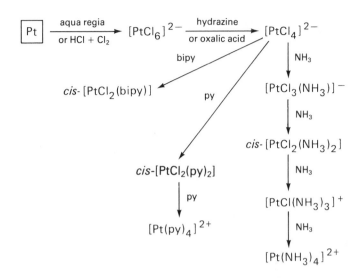

Fig. 1.5 — Synthetic routes to some platinum(II) complexes, from platinum metal.

platinum(IV) complex $[PtCl_6]^{2-}$. Addition of potassium chloride gives bright yellow $K_2[PtCl_6]$, which can be used as a source of platinum(IV) complexes. Mild reduction of $[PtCl_6]^{2-}$, for instance by hydrazine or oxalate, gives $[PtCl_4]^{2-}$. This anion too can be conveniently isolated as its potassium salt. The resulting orange-red $K_2[PtCl_4]$ is the starting material for the preparation of a whole range of platinum(II) complexes by appropriate substitution reactions, of which a very small number are shown in Fig. 1.5.

Ligand replacement reactions are also of great value in the preparation of cobalt(III) and chromium(III) complexes. Thus once $[CoCl(NH_3)_5]^{2+}$, $[Co(CO_3)(NH_3)_4]^+$, or $[Co(CO_3)_3]^{3-}$ have been prepared (see above), they can be used as starting materials for the preparation of series of other complexes (Fig. 1.6).

1.5 MODEL FOR IONS IN SOLUTION

Chapters 2 to 4 deal with many fundamental aspects of the chemistry of simple ions in solution. Before embarking on these chapters it seems sensible to define a model for an ion in solvent, and to define the terms to be used. For reasons which will emerge in the course of the following paragraphs, matters are not as clear-cut as they might be, and different people have slightly different ideas of just what is meant by the models and terms generally used. The model used here for aqueous solutions is shown in Fig. 1.7. Both cations and anions have primarily solvation shells consisting of solvent molecules interacting directly with the ions. This is region A in the diagram. Beyond these primary solvation shells are secondary solvation shells marked B. Here the solvent molecules are not in contact with the ions, but are influenced by their proximity, mainly through interaction of their dipoles with the electrostatic fields of the ions. At a sufficient distance from the ion its perturbing effect on the solvent can

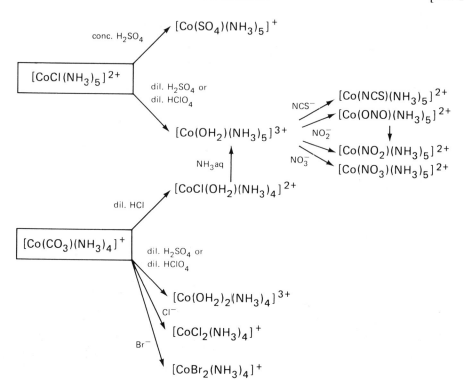

Fig. 1.6 — Synthetic routes to some cobalt(III) complexes.

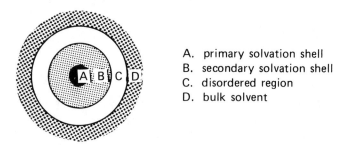

A. primary solvation shell
B. secondary solvation shell
C. disordered region
D. bulk solvent

Fig. 1.7 — The environment of an ion in aqueous solution.

be considered negligible, and such solvent molecules comprise bulk solvent. This is region D in Fig. 1.7. Now, for a solvent, such as water, with a specific three-dimensional structure, there is likely to be a mismatch between the structures of regions D and B. Hence it is necessary to consider an intermediate disordered region

C between regions D and B. Pictures similar to Fig. 1.7 will also apply to such solvents as alcohols, but simpler models will apply to solvents which lack the structural features of these polar protic solvents. One ultimate objective must be to attempt to estimate the number of solvent molecules in each of the shells. This objective is unlikely to be achieved in the foreseeable future, and indeed for some ions it may be unrealistic to define solvation shells in this manner. It should be admitted that a model such as that shown in Fig. 1.7 really only applies to very dilute solutions. Even if regions A, B, and C shown in Fig. 1.7 are each only one solvent molecule thick, that would add up to many dozen solvent molecules per ion. In a molar solution of a 1:1 electrolyte in water the ratio of water molecules to ions is rather less than 30 to 1, so the Fig. 1.7 model would apply fully only to solutions of concentrations below something like one-thousandth molar. This of course is much too dilute for many of the techniques and applications relevant to ionic solutions. Therefore is should be borne in mind that the solutions under discussion are sometimes far from ideal. They may be neither ideal in the sense of the model just discussed nor in the thermodynamic sense. Nonetheless in this topic, as in so many other real situations, a degree of approximation and compromise are essential prerequisites to progress.

2

Solvation numbers

Many techniques have been brought to bear on the problem of establishing solvation numbers for ions in solution, with varying degrees of usefulness and success. Only a few of the most important methods can be dealt with here, but some idea of the range employed can be obtained from the list set out in Table 2.1. Widely different results

Table 2.1 — Methods for estimating solvation numbers for ions

Spectroscopic	Thermochemical
NMR	entropies
ultraviolet–visible	volumes
infrared–Raman	compressibilities
Transport properties	Other methods
transport numbers	isotopic dilution
ionic conductivities	dielectric constants
mobilities	
viscosities	

have been obtained from different methods. In many cases divergences arise from difficulties in interpreting results, or in dividing results from salts into the components for their constituent ions. In other cases differences arise because various methods do not always measure the same thing (see below). Discussion of many of the other difficulties requires detailed knowledge of the methods and of the interpretation of their experimental results. Several suitable references are given in the Further Reading section at the back of this book for readers wishing to investigate these matters further.

2.1 NMR SPECTROSCOPY

Undoubtedly the most informative technique has been NMR spectroscopy, but there are several severe restrictions on its usefulness in obtaining solvation numbers for ions. In the first place one can only obtain suitable spectra for diamagnetic and for certain paramagnetic solvento-ions, depending on nuclear relaxation properties. ^{1}H NMR signals can easily be obtained for water in the vicinity of, e.g., Ni^{2+}, Co^{2+}, and the majority of lanthanide(III) cations, but Cu^{2+}, Mn^{2+}, and Gd^{3+} cause great line-broadening with its attendant problems (in this context—the extreme line-broadening and huge shifts caused by these nuclei are a positive asset in other contexts!). The other general limitation arises from the relatively long timescale of NMR spectroscopy. For proton NMR spectroscopy, a nucleus has to remain in a given environment for at least 10^{-4} s to be recognised as a distinct entity. This is a long time in relation to the timescale of molecular diffusion in the great majority of solvents. Indeed in the majority of ionic solutions, solvent molecules move between the various environments depicted in Fig. 1.7 many times within a period of 10^{-4} s. Hence for such solutions just one resonance will be seen for each type of nucleus in the molecule—one line for water, two for methanol—but there are a number of metal ions which bond strongly enough to solvent molecules for the residence time in the primary solvation shell to be longer, sometimes much longer, than 10^{-4} s. For such cases, two NMR peaks are observed for a given solvent nucleus, one corresponding to solvent moelcules in the primary solvation shell of the cation, the other to all the other solvent molecules. Some proton NMR spectra for such a situation are shown in Fig. 2.1. In these spectra the two peaks are labelled according to the conventional

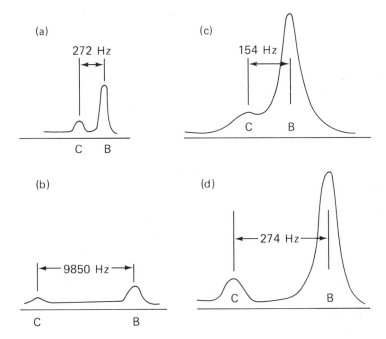

Fig. 2.1 — Proton NMR spectra of aqueous solutions of (a) gallium(III) perchlorate (1.29 M; 211 K); (b) cobalt(II) perchlorate (3.2 M; 213 K); (c) indium(III) chloride (3.81 M; 213 K); (d) indium(III) perchlorate (0.11 M in 1:5 aqueous acetone; 173 K). The signals marked C arise from the primary cation hydration shells; those marked B from the remaining water molecules in the system.

descriptions 'coordinated' and 'bulk'. 'Coordinated' refers to the primary solvation shells of the cations, while the term 'bulk' here includes not only the bulk solvent as defined earlier in connection with Fig. 1.7 but also the secondary solvation shells around both cations and anions and the primary solvation shell of the anions. Solvation numbers for cations can be obtained by straightforward arithmetic from peak areas of spectra of the type shown in Fig. 2.1 by standard integration techniques, provided that the composition of the solution is known. The arithmetic involved can be illustrated for the Mg^{2+} cation. A molar aqueous solution of this salt has a density of $1.115\,g\,cm^{-3}$. One litre of this solution will contain $985\,g$ or $985/18 = 54.7$ moles of water. If each Mg^{2+} has six waters in its primary hydration shell, then these 54.7 moles of water can be apportioned as 6.0 for Mg^{2+} cation primary hydration plus 48.7 moles of 'bulk' water. The ^1H NMR spectrum of such a solution may therefore be expected to consist of two peaks, whose areas will be in the ratio $6.0:48.7$ or $1:8.1$.

In the situation just described, and indeed in most NMR experiments designed to establish primary hydration numbers for cations, the concentration of the salt is really far too high for the model described in Chapter 1 (see section 1.5) to apply, and is much higher than anything approximating to an 'ideal solution'. Indeed at this sort of concentration it is quite possible for significant ion-pairing to occur. If the anion actually enters the primary coordination shell of the cation, then a correspondingly low value for the apparent primary hydration number will be obtained. Early estimates of five for the primary solvation number of Zn^{2+} probably arose in this way. In the early days the use of concentrations of the order of molar was necessary for this type of NMR study, and even nowadays the low sensitivity of NMR forces the use of solutions well above concentrations where 'ideal' behaviour can be assumed.

Although NMR spectra of the type shown in Fig. 2.1 can sometimes be obtained at room temperature, it is often necessary to cool solutions to slow solvent exchange sufficiently to get the two separate signals for 'coordinated' and 'bulk' solvent molecules. Such separation with decreasing temperature is illustrated in Fig. 2.2. In this respect the use of strong solutions can be a great advantage. Such solutions have considerably lower freezing points than dilute solutions, and can therefore be studied at considerably lower temperatures. Solvents such as methanol or acetone have an analogous advantage over water. In difficult cases hydration numbers have been estimated at remarkably low temperatures. Thus, for example, as shown in Fig. 2.1(d), a proton NMR spectrum has been obtained at $173\,K$ for indium(III), by the addition of acetone to a strong aqueous solution of indium perchlorate. There is a bonus for cation–solvent systems where the range of spectra from fast to slow exchange, as shown in Fig. 2.2, can be examined. That is the possibility of determining rate constants and Arrhenius parameters for solvent exchange (see Chapter 9).

The results of many 'slow-exchange' NMR studies are summarised in Table 2.2. This shows that by far the most common primary solvation number for cations is six, in water and in non-aqueous solvents. Only for very small cations, such as Be^{2+}, for the uniquely square-planar Pt^{2+} and Pd^{2+}, and for very bulky solvents at other cations, is the lower solvation number of four found. On the other hand only for the largest ions is there firm evidence for cation solvation numbers greater than six. The result of a large number of sometimes difficult experiments is the emergence of an unusually simple picture of cation primary solvation.

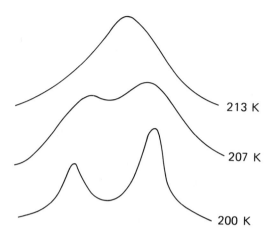

Fig. 2.2 — Variation of ^{1}H NMR spectra with temperature for magnesium perchlorate solution in aqueous acetone.

Table 2.2 — Cation solvation numbers determined from NMR peak areas

	Water	MeOH	TMP	MeCN	DMF	DMSO	liq.NH$_3$
sp-elements							
Be^{2+}	4		4		4	4	
Mg^{2+}	6	6					6
Zn^{2+}	6	6					
Al^{3+}	6	6	4		6	6	6
Ga^{3+}	6	6	4	6	6	6	
In^{3+}, Sc^{3+}			4				
Transition elements							
V^{2+}	6						
Mn^{2+}	6	6		6			
Fe^{2+}, Co^{2+}, Ni^{2+}	6	6		6	6	6	
Ti^{3+}, V^{3+}	6						
Cr^{3+}, Fe^{3+}						6	
Pd^{2+}, Pt^{2+}	4						
Lanthanides							
Ce^{3+}, Pr^{3+}, Nd^{3+}					9		
Tb^{3+} → Yb^{3+}					8		
Actinides							
Th^{4+}	9						

DMF = dimethylformamide.
DMSO = dimethyl sulphoxide.
TMP = trimethyl phosphate.

Many of the cations for which the NMR peak area method gives primary solvation numbers are included in Table 2.2. However, a large number of cations cannot appear in this table, including such important cations as those of the alkali metals, the alkaline earths, and the majority of those of the transition elements. As explained earlier, this is because solvent exchanges too rapidly with the primary solvation shell, at all accessible temperatures, for separate 'coordinated' and 'bulk' solvent NMR signals to be seen. In order to estimate solvation numbers for these cations it is necessary to use the less direct methods listed earlier (see Table 2.1). A problem common to nearly all the methods mentioned in Table 2.1 is that they give an estimate of the sum of the solvation numbers of the cation and anion of the salt examined. It is thus necessary to divide this sum into the two components. Such divisions are often difficult to make, and are sometimes carried out in a distinctly arbitrary manner. The other feature of many of the Table 2.1 approaches is that they give total solvation, primary plus secondary. For a cation whose primary solvation number is established, e.g. by NMR, this gives useful information about the secondary solvation shell, but for a fast-exchange cation, division of the total solvation number into primary and secondary shell components may cause difficulties.

2.2 ION MOVEMENT METHODS

There are a number of methods for assessing ion solvation that are based on determining the resistance to motion through the solution, thence estimating the effective volume of the moving solvated ion and from this the solvation number. The motion may be spontaneous (e.g. diffusion), mechanically engineered (e.g. viscosity), or engendered by some electrical means (e.g. conductivity). Many of these methods give the sum of cation plus anion contributions, but some do give estimates for individual ions. Thus, for instance, there are methods for measuring transference or transport numbers for ions—the Hittorf, moving boundary, and emfs of cells with transport, methods. Transport numbers are closely linked to ionic conductivities and mobilities (see physical chemistry books for details and formulae). Conductivities and mobilities, which are closely related, are of most direct relevance here, as they provide a measure of the resistance to motion through the liquid of solvated ions. This resistance reflects their effective sizes.

A selection of ionic conductivities in aqueous solution is given in Table 2.3, for simple cations and anions and for a few complex ions. Table 2.3(a) lists *equivalent* ionic conductivities, which are conductivities normalised to unit charge so that they reflect just size factors. Trends for the monatomic ions show decreasing conductivity with decreasing ionic radius. This indicates that the effective radius of the hydrated ion in fact increases as the ionic radius decreases. The particularly large $[N^nBu_4]^+$ and $[Fe(phen)_3]^{2+}$ ions have the lowest conductivities, in other words are least mobile. Table 2.3(b) lists *molar* ionic conductivities for a small selection of ions, to emphasise how much charge affects conductivities (and mobilities). Overall, molar ionic conductivities reflect charge and effective radius, i.e. the size of the hydrated ion which forms the entity which moves through the solution carrying the current.

Having established ionic mobilities, it is then possible to estimate, using fairly reasonable assumptions, diffusion coefficients and, from these, hydrodynamic radii.

Table 2.3

(a) Limiting ionic conductivities (S cm^2 equiv^{-1})^a in aqueous solution at 298.2 K

Monatomic ions							
	Li$^+$	38.7	Be^{2+}	45		F$^-$	55.4
	Na$^+$	50.1	Mg^{2+}	53.1	La^{3+} 69.7	Cl$^-$	76.4
	K$^+$	73.5	Ca^{2+}	59.5	$\downarrow$	Br$^-$	78.1
	Rb$^+$	77.8	Sr^{2+}	59.5	Yb^{3+} 65.6	I$^-$	76.8
	Cs$^+$	77.3	Ba^{2+}	63.6			

Polyatomic ions				
	NMe$_4^+$	44.9	ClO$_4^-$	67.3
	NBu$_4^+$	19.5	ReO$_4^-$	55.0
			SO$_4^{2-}$	80.0

Complex ions   [Fe(phen)$_3$]$^{2+}$ [Co(NH$_3$)$_6$]$^{3+}$ [Fe(CN)$_6$]$^{3-}$ 101
34.3   102

[Fe(CN)$_6$]$^{4-}$ 111

(b) Selected limiting ionic conductivities (S cm^2 mol^{-1}), derived from section (a) above by multiplying by n for an $n+$ or $n-$ ion

	K$^+$	73.5	Ca^{2+}	119	La^{3+}	209

[Co(NH$_3$)$_6$]$^{3+}$ 306   [Fe(CN)$_6$]$^{3-}$ 303
[Fe(CN)$_6$]$^{4-}$ 444

[P$_3$O$_{10}$]$^{5-}$ 545

^{a}S (Siemens) = reciprocal ohm or mho.

A hydrodynamic radius gives the volume of the hydrodynamic sphere—that is the sphere consisting of the ion plus its attached solvent molecules. The volume of the ion itself can be subtracted from this to get the volume of the solvent molecules. From this it is a simple matter to calculate the number of solvent molecules involved.

2.3 REVIEW OF SOLVATION NUMBERS

A very small selection of estimates of hydration numbers for 'fast exchange' cations, obtained via mobilities and by a range of other approaches, is given in Table 2.4. This includes some 'slow-exchange' cations for comparison. There are two main points to be made about the numbers in Table 2.4. The first is that for any given cation and solvent, different methods can give significantly, and occasionally dramatically, different results. The second is that the average values from kinetic methods—those which involve the actual *movement* of solvated ions, such as diffusion, viscosity, conductivity, and transport numbers—are generally much greater than six. This point is emphasised by comparing the NMR peak area values of six for Mg^{2+}, Zn^{2+}, Fe^{2+}, Al^{3+}, and Cr^{3+} with the much higher values obtained from kinetic methods. The main reason for this difference is that the ion-movement methods give an

Table 2.4 — Cation hydration numbers

	Li^+	Na^+	K^+	Cs^+	Mg^{2+}	Ca^{2+}	Ba^{2+}	Zn^{2+}	Fe^{2+}	Al^{3+}	Cr^{3+}
Transport numbers	13–22	7–13	4–6	4	12–14	8–12	3–5	10–13			
Mobilities	3–21	2–10	5–7		10–13	7–11	5–9	10–13	10–13		
Conductivities	2–3	2–4		6	8	8	8				
Diffusion	5	3	1	1	9	9	8	11	12	13	17
Entropies	5	4	3	3	13	10	8	12	12	21	
Compressibilities	3	4	3							31	
Activity coefficients					5	4	3		12	12	
Cf. NMR peak areas					6			6	6	6	6

estimate of the average number of solvent molecules which move with each ion, which clearly includes secondary as well as primary solvation shells. A general impression emerges that the secondary hydration shell of Mg^{2+} and the 2+ ions of the first row of the d-block elements contains some six to eight water molecules. It is thus effectively only one molecule thick. Comparison of the values for Al^{3+} with those for divalent cations indicates a much bigger secondary hydration shell for Al^{3+}, which is as one would expect from its much stronger electrostatic field. Table 2.4 includes several sequences where decreasing electrostatic field arising from increasing cation radius appears to result in decreasing secondary hydration, as on descending a Periodic Table Group. The variability of secondary hydration numbers for a given cation reflects experimental difficulties, problems in splitting total hydration numbers for salts into their ionic components, and the fact that, particularly for ion-movement methods, hydration numbers may actually depend on the technique used. An ion may well be accompanied by a somewhat larger hydration shell when it moves spontaneously (i.e. diffuses) than when it is made to move by the application of a mechanical or electrical force (in viscosity or conductivity experiments, for example). The data in Table 2.4 are restricted to hydration numbers, but similar trends may be expected for solvation in other solvents. However, the rather larger size of most organic solvent molecules will tend to give somewhat lower solvation numbers, especially for the secondary solvation shell.

As stated in the NMR section, the majority of metal ions, particularly those of d-block elements, have a primary solvation number of six, in water and in organic solvents. The particularly small Be^{2+} cation and the square-planar Pd^{2+} and Pt^{2+} cations have primary solvation numbers of four, while very large organic molecules may give solvation numbers of 4 for some normally octahedral d-block cations (e.g. $[Co(hmpa)_4]^{2+}$). The situation in relation to primary hydration numbers for the 3+ and 4+ ions of the large f-block elements is not entirely clear. There is some evidence to suggest that the 3+ ions of the earlier lanthanides have a hydration number of 9, of the later and slightly smaller lanthanides 8. NMR and diffraction studies of aqueous solutions containing 4+ actinide cations (thorium; uranium) indicate primary hydration numbers of 8 or 9. There is an almost complete absence of information on secondary hydration numbers for lanthanide and actinide cations.

Finally it should be stated that varying temperature or pressure very rarely has any effect on primary solvation numbers of cations. In the case of Co^{2+}, where tetrahedral geometry is relatively little disfavoured in relation to octahedral, there is some evidence for the establishment of a $[Co(OH_2)_6]^{2+} \rightleftharpoons [Co(OH_2)_4]^{2+} + 2H_2O$ equilibrium at elevated temperatures. There is both NMR and ultraviolet-visible spectroscopic evidence for an increase in solvation number from 8 to 9 for Yb^{3+} in DMF on increasing the pressure.

3

Ion–solvent distances

3.1 X-RAY DIFFRACTION BY SOLUTIONS

The next fundamental aspect of the nature of metal ions in solution is the distance between the metal atom and solvating solvent molecules, or more precisely between the metal atom and the atom in the solvent molecule which is bonded to the metal ion. In aqueous and alcoholic solutions this is, of course, the oxygen atom of the solvent molecule. In recent years the development of X-ray diffraction techniques and their application to this problem have begun to give accurate values for these metal-ion-to-primary-solvent distances, and are beginning to give information on distances to secondary solvent molecules.

The use of X-ray diffraction methods to obtain structural information on crystalline materials is very familiar. The application of such methods to solutions may be less familiar, though in fact such an application was first reported way back in 1929. X-ray studies of solutions are technically little more difficult than those of crystalline solids, but the processing of the experimental results — obtaining and interpreting the radial distribution functions — is considerably more taxing. Before dealing with the method of estimation and with the results obtained, it might be helpful to introduce radial distribution functions and to indicate their relation to the geometries of solvated ions. The basic concepts outlined in the following section are also relevant to related experimental approaches such as neutron diffraction, and X-ray and neutron scattering — more rarely used than X-ray diffraction at present, but of rapidly increasing usefulness and importance.

3.2 RADIAL DISTRIBUTION FUNCTIONS

We shall start this discussion of radial distribution functions with a consideration of some ordered and disordered arrangements in two dimensions. Both visualisation and drawing are easier in two dimensions than in three, but the principles are the same. The left-hand side of Fig. 3.1 shows representations of ordered, disordered,

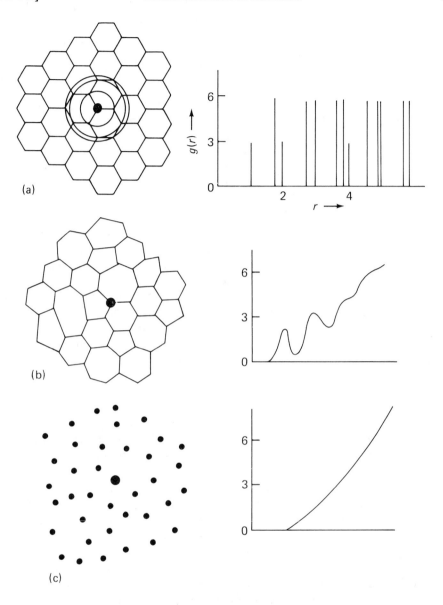

Fig. 3.1 — Two-dimensional radial distribution functions; r in (a) is the hexagon side length.

and random arrangements. The top arrangement is simply part of an infinite array of regular hexagons, as in a layer of graphite. In the middle diagram all the vertices are three-connected as in the top diagram, but now the vertices make up pentagons, hexagons, and heptagons, at random. Finally the bottom diagram shows a random arrangement of points, as in a two-dimensional gas. The right-hand side of Fig. 3.1 shows the respective radial distribution functions, $g(r)$. A radial distribution function represents the probability of finding a particle at a given distance from a set point —

in the present discussion the probability of finding a solvent molecule at a given distance from the ion being solvated. This is done in terms of the number of points at given distances from that central point. In fact the radial distribution functions in Fig. 3.1 are averaged over the whole of two-dimensional space. In the case of the regular hexagonal array at the top, every vertex has three nearest neighbour vertices along one side of the hexagons. Every vertex then has six next nearest neighbours. There are two equivalent neighbours in what might be called the meta-positions of the three hexagons sharing the central vertex. The next-nearest neighbours are three in number, diagonally across the hexagons, in other words in the para positions. These numbers and distances are appropriately displayed in the top radial distribution function, which includes the relation of all the vertices in the Fig. 3.1 hexagons diagram to the central vertex.

Now, we turn to the middle arrangement, of three-connected pentagons, hexagons, and heptagons. Here the number of nearest neighbour vertices is again always three, but the geometric consequence of packing three types of polygon together is that all sides can no longer be quite equal. So the first peak in the radial distribution function is no longer sharp but somewhat broadened, to include the actual range of nearest-vertex distances. The range of next-nearest vertex distances will obviously be greater, giving a second broader peak. Very soon the peaks blur together, as shown in the middle plot of Fig. 3.1. Finally, the radial distribution for the two-dimensional gas shows no peaks, as the distances of the various points from a given point are random. The plot curves smoothly upwards with distance, as the chances of meeting a point at a given distance increase as the circumference of a circle increases with its radius.

Now, moving on to the real world, Fig. 3.2 shows the three-dimensional array of ions in sodium chloride and an impression of the random arrangement of octahedral solvento-ions in solution. Opposite these representations are shown the respective radial distribution functions. In the case of the ions in solution both the radial distribution function for one ion and, below this, the average radial distribution function for the whole array of ions, are given. With the sodium chloride structure, the regular geometrical arrangement gives a radial distribution function with fixed numbers of neighbours at clearly defined distances, as in the two-dimensional case of the regular hexagonal array. This sequence is the same as that used to generate the Madelung constant series for this structure. In the case of solvated ions in solution, each ion has six nearest neighbours at a clearly defined distance. The next shortest regular distance is that between *cis*-solvent molecules in the octahedral primary solvation shells. For several solvents this approximates to nearest neighbour solvent–solvent distances in bulk solvent, so the finite range of distances here begins to blur the radial distribution function. With increasing distance this soon degenerates into the smooth curve characteristic of the two-dimensional gas of Fig. 3.1. Comparison of the two parts of Fig. 3.2 emphasises that, whereas in crystals there is both short-range and long-range order, in solutions there is only short-range order.

3.3   RESULTS OF DIFFRACTION AND SCATTERING STUDIES

The most prominent features in radial distribution functions for ions in solution are thus the nearest-neighbour peaks, from which the metal to solvent distance for the

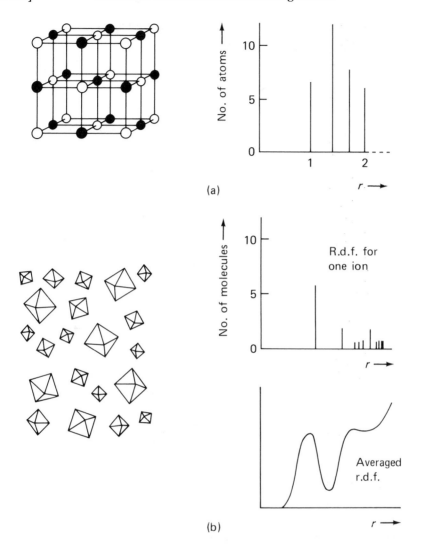

Fig. 3.2 — Three-dimensional radial distribution functions (r.d.f.). (a) Sodium chloride structure. (b) Hexasolvento-ions in solution.

primary solvation shell can be obtained. Table 3.1 shows metal to oxygen distances determined for metal ions in aqueous solution by X-ray and neutron techniques, both diffraction and scattering in type. These distances parallel crystal ionic radii, and are very similar to those reported for analogous crystal hydrates. Where cation–oxygen distances in crystal hydrates cover a span of values, as in hydrated calcium or potassium salts, then the solution results give a more satisfactory estimate of the 'real' ion–solvent distance, unaffected by packing complications. Distances of 2.10 and 1.98 Å for $[Fe(OH_2)_6]^{2+}$ and $[Fe(OH)_2)_6]^{3+}$ respectively, and of 2.11 and 2.03 Å for $[Ru(OH_2)_6]^{2+}$ and $[Ru(OH_2)_6]^{3+}$ respectively, show the expected significant shortening of the ion–water distances in the higher oxidation state in both cases.

Table 3.1 — Metal ion to oxygen distances (Å) in aqua-cations

	Aqueous solution			Crystal hydrates	Ionic[a] radii
	X-ray diffrn.	EXAFS	Neutron diffrn.		
Li^+			1.90–1.95	1.93–1.98	0.88
Na^+	2.40		2.50	2.35–2.52	1.16
K^+	2.87–2.92		2.70	2.67–3.22	1.52
Ag^+	2.43	2.31–2.36			1.29
Mg^{2+}	2.10			2.01–2.14	0.86
Ca^{2+}	2.40		2.39–2.46	2.30–2.49	1.14
Cr^{2+}	2.07; 2.30				
Mn^{2+}	2.20	2.18		2.00–2.18	0.96
Fe^{2+}	2.12	2.10		1.99–2.08	0.91
Co^{2+}	2.08	2.05		1.93–2.12	0.88
Ni^{2+}	2.04	2.05–2.07	2.07–2.10	2.02–2.11	0.84
Cu^{2+b}	1.94; 2.43	1.95;[c]	1.95–2.05	1.93–2.00[d]	0.87
Zn^{2+}	2.08–2.17	1.94		2.08–2.14	0.89
Cd^{2+}	2.31			2.24–2.31	1.09
Hg^{2+}	2.41			2.24–2.34	1.16
Al^{3+}	1.87–1.90			1.87	0.67
In^{3+}	2.15			2.23	0.93
Cr^{3+}	1.94	1.98		2.02	0.76
Fe^{3+}	2.05	1.99		2.09–2.20	0.79
Rh^{3+}	2.04–2.07			2.09–2.10	0.81
Ce^{3+}	2.55			2.48–2.60	1.19
Nd^{3+}	2.51		2.48	2.47–2.51	1.14
Dy^{3+}	2.40		2.37	2.38	1.05
U^{4+}	2.42			2.36	1.14

[a]For six coordination.
[b]Cu^{2+} has a tetragonal environment in aqueous solution (Fig. 3.7).
[c]It has proved difficult to locate the axial water molecules by EXAFS.
[d]Cu^{2+} aq in crystalline hydrates is often less symmetrical even than tetragonal.

Of the X-ray and neutron techniques used to obtain ion–solvent distances in solution, X-ray diffraction is probably the least difficult to use. However, neutron diffraction offers the great advantage over X-ray diffraction of locating hydrogen, or rather deuterium, atoms and thus of giving information about the orientation of solvating water molecules with respect to the ions solvated. Fig. 3.3 shows the radial distribution function about a metal ion in D_2O. The two main peaks correspond to M–O and M–D separations. As both distances are known, simple trigonometry gives the orientation of the solvating D_2O molecule, as shown in M=Li in Fig. 3.4. In this respect neutron diffraction is more informative than X-ray diffraction. However, the equipment and isotopes needed for neutron diffraction studies of electrolyte solutions are extremely expensive, so results are available for only relatively few ions as yet. The appropriate column in Table 3.1 gives a good idea of the relative scarcity of neutron diffraction data.

In recent years a variety of new applications of X-ray and neutron techniques have begun to yield information relevant to the nature of ions in solution. The most important of these is EXAFS — Extended X-ray Absorption Fine Structure. This technique is particularly useful for probing the immediate environment of a particular element in a poorly ordered system such as a solution, where order is short-range.

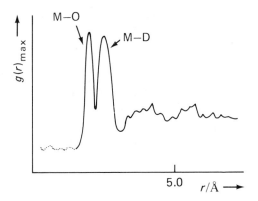

Fig. 3.3 — A typical radial distribution function for a metal ion in D_2O, as obtained from neutron diffraction.

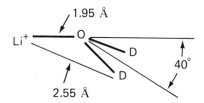

Fig. 3.4 — The geometry of the D_2O-Li^+ solvate moiety.

It offers the advantage over X-ray diffraction of providing information specifically about the surroundings of the element whose absorption edge has been selected — X-ray diffraction gives information about near neighbour interactions between all elements (except H) in the system. EXAFS has proved successful in obtaining detailed information about, for example, primary and secondary solvation shells of

Cu^{2+} in aqueous solution, about μ-hydroxo-dinuclear cations of the $M\begin{smallmatrix}OH\\OH\end{smallmatrix}M$

type (Chapter 5), and about ternary aqua-ligand complexes of a variety of metal ions. A similar range of information is becoming available from X-ray scattering and neutron scattering, e.g. from WAXS (Wide Angle X-ray Scattering), and SANS (Small Angle Neutron Scattering). Such studies give estimates for solvation numbers and for ion–solvent distances, and some idea of the nature of secondary solvation shells. In favourable cases the solvent environment can be probed to 20 or 30 Å from a metal ion, though EXAFS probing is limited to rather shorter distances. EXAFS is thus particularly valuable for studying the immediate environment of ions, particularly primary solvation shells (see Table 3.1). These and related developments are still in their infancy, but may be expected to feature prominently in the study of solvated ions and complexes in the near future.

Both X-ray diffraction and newer techniques such as EXAFS have proved useful in probing solvation geometries more complicated than normal simple regular octahedral or tetrahedral. Thus, for example, the tetragonal distortion of primary coordination shells around Cu^{2+}, well documented for complexes in the solid state, can be seen both in radial distribution functions and in EXAFS studies (see the previous paragraph) of copper(II) salts in aqueous solution (see Table 3.1). Here two metal to oxygen peaks can be distinguished, at 1.94 and 2.4 Å, corresponding to the four equatorial and two axial water molecules respectively, as indicated in Fig. 3.5

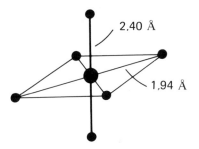

Fig. 3.5 — Tetragonal distortion in $[Cu(OH_2)_6]^{2+}$.

and Table 3.1. Similarly, chromium–oxygen distances of 2.07 and 2.30 Å have been established by EXAFS in aqueous solutions of the other important Jahn–Teller distorted aqua-ion, $[Cr(OH_2)_6]^{2+}$ (d^4).

It is sometimes possible to deduce the geometry of primary solvation shells from radial distribution functions, by comparing, e.g., oxygen–oxygen distances for pairs of water molecules in a primary hydration sphere with the metal–oxygen distance. In octahedral $[Fe(OH_2)_6]^{2+}$, the oxygen–oxygen distances (octahedron edges) are 1.41 times the iron–oxygen distance; in tetrahedral $[FeCl_4]^-$, the chlorine–chlorine distances (tetrahedron edge) are 1.73 times the iron–chlorine distance (see Fig. 3.6).

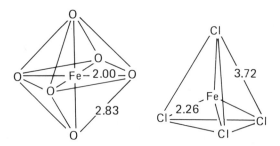

Fig. 3.6 — Primary coordination sphere geometries.

Tetrahedral $[Fe(OH_2)_4]^{2+}$ would have oxygen–oxygen distances of 3.46 Å, considerably longer than 2.83 Å in octahedral $[Fe(OH_2)_6]^{2+}$. In general, the radial distribution function for an octahedral aqua-ion $[M(OH_2)_6]^{n+}$ with M–O=r will include peaks at r, $1.41r$, $2r$,. . .; for a tetrahedral aqua-ion peaks will be at r, $1.73r$,. . . . These cases, and the expectations for square-planar (e.g. Pd^{2+} aq) and linear (in principle possible for heavy d^9 and d^{10} ions, e.g. Ag^+ or Hg^{2+}) solvento-cations, are depicted in Fig. 3.7.

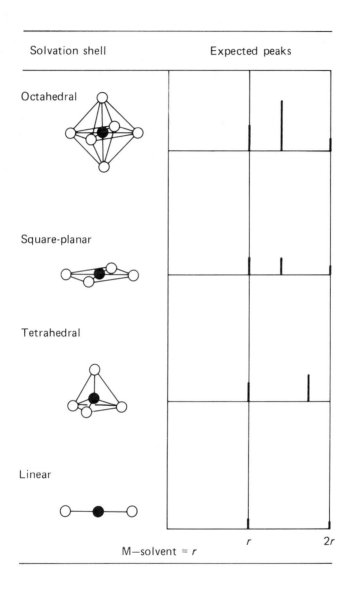

Fig. 3.7 — Expected peaks in radial distribution functions for various solvation shell geometries.

Recently the peak integration method, familiar from NMR spectroscopy, has been applied to radial distributions determined by X-ray or neutron diffraction methods. For ions such as Ni^{2+} in water, the expected hydration number of six has been obtained. For U^{4+}, for which no satisfactory NMR peak area estimate of hydration number exists, X-ray diffraction peak integration has indicated the presence of nine water molecules in the primary hydration shell.

Now that metal–oxygen distances have been established for a large number of cations in aqueous solution, attention is being turned to the more difficult problem of probing the constitution and geometry of secondary solvation shells. Diffraction methods generally involve the use of fairly concentrated solutions, so there may well be one or more anions present in the secondary solvation shell of a metal ion. In the absence of significant ion-pairing, it is possible to come to some conclusions about the nature of secondary hydration shells for certain cations. Thus X-ray diffraction radial distribution functions for aqueous solutions containing Mg^{2+}, Al^{3+}, Cr^{3+}, or Fe^{3+} all suggest that there are water molecules in a secondary solvation shell at a M^{n+}-O distance of 4.1 to 4.2 Å. This seems reasonable, as do estimates of 14 to 18 for total hydration numbers (see Table 2.4). There is a hint of the secondary solvation shell in the radial distribution functions shown in Fig. 3.2(b) and Fig. 3.3, but these metal-ion-to-secondary-solvent distances and estimates of secondary hydration numbers are more readily derived from differential plots.

4

Ion–solvent interactions

The third fundamental aspect of the nature of ions in solution is the strength of bonding between an ion and the solvent molecules in its primary solvation shell. This has been investigated in a variety of ways, mainly thermodynamic and spectroscopic. The gauging of ion–solvent bond strengths through estimates of such parameters as ion solvation enthalpies will be dealt with in the thermodynamics section (section 4.3). Before this, the two most useful spectroscopic approaches, ultraviolet–visible and infrared–Raman, will be discussed in sections 4.1 and 4.2. Applications of ultraviolet–visible spectroscopy to ion–solvent interactions are generally restricted to metal ions of the d-block elements, whereas infrared–Raman spectroscopy is, of course, of considerably wider applicability.

4.1 ULTRAVIOLET–VISIBLE SPECTROSCOPY

For the majority of transition metal ions, strengths of ion–solvent interactions can be estimated from ultraviolet–visible absorption spectra with the aid of crystal field or molecular orbital theory at an appropriate level of sophistication. The simplest application is to ions with a single d electron, for example Ti^{3+}. An octahedral field, as in a hexasolventocation, splits the d levels into t_{2g} and e_g levels in the familiar pattern shown in Fig. 4.1 (left). The corresponding term splitting is shown in Fig. 4.1

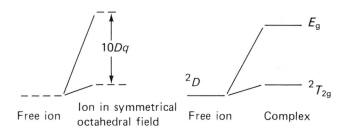

Fig. 4.1 — Effect of ligand (solvent) coordination on the d levels and ^2D term of a transition metal cation.

(right). The frequency corresponding to the single d electron transition between t_{2g} and e_g levels reflects the difference in energy between these levels, and hence the strength of interaction between solvent and cation. This frequency can be obtained from the ultraviolet–visible absorption of an aqueous solution of a titanium(III) salt, of a non-complexing anion (Fig. 4.2). The energy is calculated from the frequency by

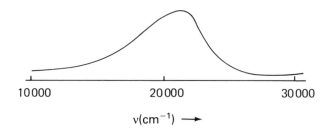

Fig. 4.2 — Ultraviolet–visible spectrum of Ti^{3+}aq in aqueous solution.

the familar equation $E = h\nu$. Table 4.1 lists the maximum absorption frequencies of

Table 4.1 — Wavenumbers of maximum absorption for Ti^{3+} in water and in alcohols

Solvent	$\nu_{max} (cm^{-1})$
Water	20080
Methanol	16850
Ethanol	16750

titanium(III) solutions in water, methanol, and ethanol. This indicate that methanol and ethanol interact with Ti^{3+} to approximately equal extents, but that water interacts rather more strongly.

Unfortunately even titanium(III) has its drawbacks. Chemically it is strongly reducing, which restricts the range of solvents which can be studied. Also its d^1 configuration results in small Jahn–Teller distortions from octahedral symmetry, so that the ultraviolet–visible absorption spectra of titanium(III) solutions consist of a broad *un*symmetrical peak rather than a sharp symmetrical peak. This is apparent in the spectrum reproduced in Fig. 4.2 above. In principle copper(II) should be as useful as titanium(III) in assessing ion–solvent interactions. This is because the d^9 configuration of copper(II), with one hole, is in many respects equivalent to the d^1 configuration of titanium(III), with its one electron. However, distortions from octahedral symmetry are generally more marked for copper(II) than for titanium(III). Such distortion were mentioned earlier in the section on X-ray diffraction studies of aqueous solutions—the geometry of hexa-aquo-copper(II) is shown in Fig 3.5.

For transition metal cations with configurations d^2 to d^8 inclusive, the term-splitting diagrams are more complicated than for the d^1 or d^9 systems. The Russell–Saunders treatment of these cases can be found in the Further Reading section at the end of the book. The spectrum and term-splitting diagram for the d^8 ion Ni^{2+} are illustrated in Fig. 4.3, which shows the greater complexity than in the d^1 or

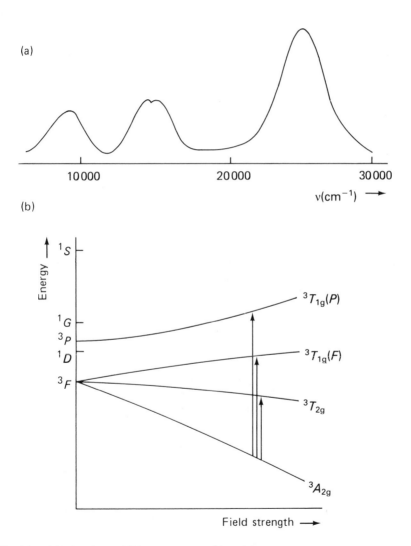

Fig. 4.3 — (a) Ultraviolet–visible spectrum of Ni^{2+}aq; (b) the three transitions involved in the absorption of a d^8 cation, e.g. Ni^{2+}aq, in an octahedral ligand field.

d^9 systems. Nonetheless, values for the crystal field parameter Δ or $10Dq$ can readily be obtained for d^2 to d^8 ions in ranges of solvents from the bands observed in their ultraviolet–visible spectra. The dependence of crystal field splitting on cation is shown, for aqueous media, in Table 4.2. The dependence of crystal field splitting on

Table 4.2 — Dependence of the Crystal Field splitting parameters, $Dq\,(\mathrm{cm}^{-1})$, on the nature of the solvent

	Co^{2+}	Ni^{2+}	Cr^{3+}
NN-Dimethylacetamide	805	796	1538
Dimethyl sulphoxide	840	777	1587
Trimethyl phosphate	975	796	1628
Ethanol			1668
Methanol		850	1695
NN-Dimethylformamide		850	1710
Water	1000	875	1720
Acetonitrile	1022	1026	
Liquid ammonia	1010	1080	2155

the nature of the solvent is illustrated by the selected values for the three cations Co^{2+}, Ni^{2+}, and Cr^{3+} in Table 4.3. Finally, crystal field effects for solvents are placed

Table 4.3 — Values of the Crystal Field splitting parameter, $Dq\,(\mathrm{cm}^{-1})$, for transition metal cations in aqueous solution

d^1			Ti^{3+}	2030		
d^2			V^{3+}	1860		
d^3	V^{2+}	1230	Cr^{3+}	1700	Mo^{3+}	2630
d^4	Cr^{2+}	1410	Mn^{3+}	2000		
d^5	Mn^{2+}	850	Fe^{3+}	1400		
d^6	Fe^{2+}	1000	Co^{3+}	1820	Rh^{3+}	~2700
d^7	Co^{2+}	930				
d^8	Ni^{2+}	890				
d^9	Cu^{2+}	1200				

in the context of crystal field effects for ligands in general in Fig. 4.4. This figure, and

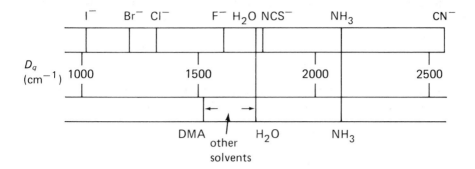

Fig. 4.4 — Correlation of solvent and ligand spectrochemical series for chromium(III).

the preceding Tables 4.2 and 4.3, show that $10Dq$ values, and thus strengths of metal ion–solvent interactions, are determined by such properties of the cations as size, charge, and d electron configuration, and by such properties of the solvents as basicity and σ-donor ability. It is interesting how much smaller the range of Dq values is for solvents than for ligands in general.

4.2 INFRARED AND RAMAN SPECTROSCOPY

To turn to a different region of the spectrum, the second conceptually fairly straightforward approach to estimating relative strengths of ion–solvent interactions is by vibrational, in other words infrared and Raman, spectroscopy. Here one uses the simple idea that, particularly in a sequence of rather similar species, stronger bonding may be correlated with higher vibrational frequencies. Often the dissolution of a salt not only affects the solvent's vibrational frequencies but also gives rise to new absorptions or peaks which can be assigned to metal–solvent vibrations. The former effect is illustrated in Fig. 4.5, which gives an idea of how the attachment of a heavy

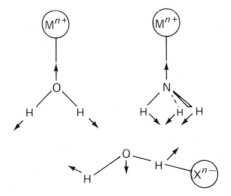

Fig. 4.5 — The relation of modes of vibration of solvent molecules to their coordination to cations and anions.

and charged metal ion may affect the normal modes of vibration indicated. The relative effects of a range of cations on the hydrogen–oxygen stretch, v_1, for solvent water are:

$$Ce^{4+} > Al^{3+} > Ba^{2+} > Mg^{2+} \sim Li^+ > Na^+ \sim K^+$$

The order given is approximately consistent with the usual combination of charge and radius effects. However, Table 4.4 shows that the situation is more complicated

Table 4.4 — Perturbation of solvent molecule vibrations by metal ions—effects of coordination on acetonitrile stretching frequencies

	Δv(C–C) (cm^{-1})	Δv(C≡N) (cm^{-1})
Li^+	+15	+22
Na^+	+8	+12
K^+	+6	+10
Rb^+	+5	+9
Cs^+		+9
Mg^{2+}	+20	+37
Co^{2+}	+18	+35
$Ru(NH_3)_5^{2+}$		−15
$Ru(NH_3)_5^{3+}$		+32

in the case of acetonitrile. Whereas in water σ-effects are predominant, in acetonitrile one has to consider not only the primarily electrostatic effects of cation charge and radius, but also the balance between σ and π bonding between cation and acetonitrile, and the transmission of these effects to the carbon–nitrogen triple bond in the coordinated acetonitrile.

Although these effects of cations on solvent vibrational frequencies are informative, the new bands assignable to cation–solvent vibrations are the key to assessing relative strengths of cation–solvent bonding. A selection of vibrational frequencies for such cation–solvent vibrations is presented in Table 4.5. There is a marked decrease in frequency on descending a Periodic Table Group, for example from lithium to caesium, from beryllium to barium, or from aluminium to indium. Such decreases may be attributed both to decreasing strength of cation–solvent bonding and, to some extent, to increasing cation mass. Looking along a row and comparing the frequencies for a selection of solvents for a given cation reveals much less variation in the values. Nonetheless, differences between water and liquid ammonia, between most pairs of organic solvents, and even between H_2O and D_2O, are significant.

Table 4.5 — The dependence of vibrational frequencies for coordinated solvent molecules (cm^{-1}) on the nature of the cation and of the solvent

	Water		Liquid ammonia	DMSO	Acetone	THF	Propylene carbonate
	H_2O	D_2O					
Li^+				429	420–425	412–413	397
Na^+			194	195–206	195–196	175–198	186
K^+				153–154	148	142	144
Cs^+				109–118			112
Ag^+			260–263				
Be^{2+}	530–543		485				
Mg^{2+}	360–365	344	328–330				
Ba^{2+}			215				
Zn^{2+}	385–400		435–440				
Al^{3+}	520–526	503					
Ga^{3+}	475		477				
In^{3+}	400		440				

4.3 THERMOCHEMISTRY OF ION SOLVATION

The importance of ion hydration enthalpies in determining solubilities of salts was emphasised at the very beginning of this text (Introduction and Fig. 1.1). There it was shown, using the specific example of sodium chloride, that enthalpies of solution generally represent small differences between large lattice enthalpies and large ion hydration enthalpies. The main topics of this section are the dependence of ion hydration enthalpies on the nature of ions and the indications of strengths of ion–solvent interactions given by ion solvation enthalpies (cf. spectroscopic indications of strengths of ion–solvent interactions discussed in sections 4.1 and 4.2 above). However, a few sentences on the derivation of ion solvation enthalpies is required before their values are discussed.

As is apparent from Fig. 1.1(a), reproduced again here as Fig. 4.6(a), it is relatively straightforward to estimate the sum of the hydration enthalpies of the pair of ions which constitute a salt. The enthalpy of solution can be measured directly in a calorimeter if the salt is reasonably soluble. If the salt is sparingly soluble, then a satisfactory estimate for its enthalpy of solution can usually be obtained from the temperature-dependence of its solubility. As the enthalpy of solution is generally the small difference between two large quantities, the approximations inherent in this van't Hoff approach are negligible in the present context. The lattice enthalpy of a salt containing monatomic ions can be calculated with reasonable precision; the lattice enthalpies of salts containing complex cations or anions are much more difficult to calculate satisfactorily, due to uncertainties in charge distribution within polyatomic ions. Sums of ion hydration enthalpies and the measured enthalpies of

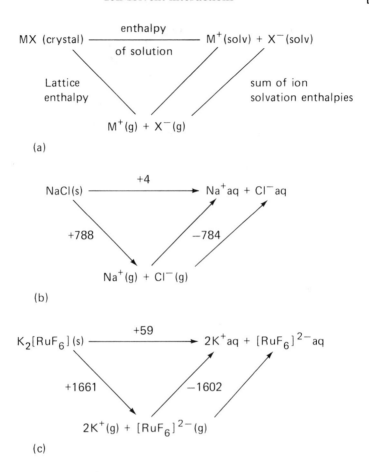

Fig. 4.6 — Interrelation of solution, solvation, and lattice enthalpies (a) in general; (b) for sodium chloride; (c) for potassium hexafluororuthenate(IV). All enthalpies are in kJ mol^{-1}.

solution and calculated lattice enthalpies from which they are derived are shown, for the specific examples of sodium chloride (cf. Fig. 1.1(b)) and potassium hexafluororuthenate(IV), in Fig. 4.6(b) and (c).

By this approach, an extensive set of sums of ion hydration enthalpies can be built up. Such a set of values can be shown to be self-consistent, but no amount of arithmetical manipulation allows the extraction of hydration enthalpies for individual ions. To obtain these single ion values one has to introduce an extra-thermodynamic assumption. The assumption of equal values for K^+ and Cl^- has often been used in various contexts over many decades. In view of the significant differences in ionic radii (Table 4.6 suggests $K^+ = F^-$ or $Cs^+ = Cl^-$ might be better) and in the geometry of hydration (Fig. 4.7), this seems fairly unattractive here. The currently popular single ion assumption of equality of $AsPh_4^+$ or PPh_4^+ and BPh_4^- is better in view of the large size and equality of radii, though Fig. 4.7 still applies. Also

Table 4.6 — Ionic radii (in Å; Shannon and Prewitt values for six coordination)

Na^+	1.02	F^-	1.33
K^+	1.38	Cl^-	1.81
Rb^+	1.49	Br^-	1.96
Cs^+	1.70	I^-	2.20

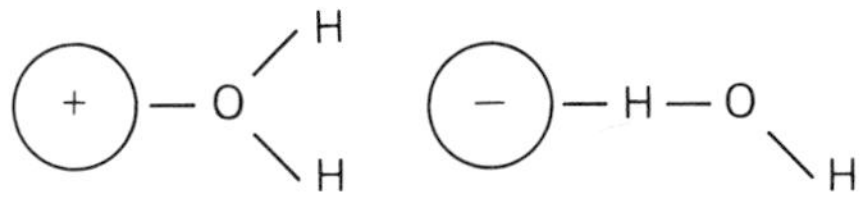

Fig. 4.7—Geometrical relations between a solvating water molecule and a cation and an anion.

the very low solubility of $[Ph_4As][BPh_4]$ and of its phosphorus analogue precludes direct calorimetry, and lattice enthalpy calculations are complicated by the polyatomic nature of these ions (see above). In practice the best approach has proved to be that of obtaining a good estimate for the hydration enthalpy of the proton†. In essence it involves the extrapolation of enthalpy data for a series of compounds HX to the limit when X^- is infinitely large and negligibly solvated. The hydration enthalpy of H^+ can then be taken as the hydration enthalpy of HX. Once the value for $\Delta H_{hydr}(H^+)$ has been fixed $(-1091 \text{ kJ mol}^{-1})$ then values for, e.g., $\Delta H_{hydr}(X^-)$ for X = Cl, Br, I, can be obtained from enthalpies of solution of HX and ancillary thermodynamic data, then $\Delta H_{hydr}(M^{n+})$ from enthalpies of solution of halides MX_n, and so on.

Enthalpies of hydration for a range of metal ions and for a selection of anions are listed in Table 4.7. There is a general overall correlation with ionic charges and radii but with a number of significant deviations from such simple electrostatic control. The dependence of hydration enthalpy on ionic radius, at constant charge, can readily be seen in Table 4.7 by looking down the members of a Periodic Table Group such as $Li^+ \rightarrow Cs^+$ or $Be^{2+} \rightarrow Ra^{2+}$. The dependence on charge is qualitatively obvious on looking across Table 4.7, but is better demonstrated in the manner shown in Table 4.8. In this latter table an attempt has been made to show the effect of changing the charge but keeping the ionic radius approximately constant. This is more satisfactory than simply noting the trend across the Periodic Table, for a sequence such as $K^+ \rightarrow Ca^{2+} \rightarrow Sc^{3+}$ involves a marked decrease in ionic radius as the charge increases. Table 4.7 shows that hydration enthalpies of anions follow the

---

† This can be complicated — see H. F. Halliwell and S. C. Nyburg, *Trans. Faraday Soc.*, **59**, 1126 (1963) for details of the best method.

Table 4.7 — Hydration enthalpies (kJ mol^{-1}) for cations and anions

Li$^+$	-515	Be^{2+}	-2487					
Na$^+$	-405	Mg^{2+}	-1922	Al^{3+}	-4660			
K$^+$	-321	Ca^{2+}	-1592	Sc^{3+}	-3960			
Rb$^+$	-296	Sr^{2+}	-1445	Y^{3+}	-3620			
Cs$^+$	-263	Ba^{2+}	-1304	La^{3+}	-3283	Ce^{4+}	-6490
						Th^{4+}	-4220

F$^-$	-503	CN$^-$	-365		
Cl$^-$	-369	NCS$^-$	-328		
Br$^-$	-336	NO$_3^-$	-328		
I$^-$	-298	ClO$_3^-$	-307		
		ClO$_4^-$	-244	SO$_4^{2-}$	-1145

Table 4.8 — Dependence of cation hydration enthalpy on charge at constant ionic radius

Cation	Radius (Å)	ΔH_{hydr} (kJ mol^{-1})
Na$^+$	1.16	-405
Ca^{2+}	1.14	-1592
Nd^{3+}	1.14	-3440
Pu^{3+}	1.14	-3440

same pattern as those for anions. The different modes of interaction (Fig. 4.7) result in the hydration enthalpies of anions and cations of similar radius being significantly different (see, e.g., K$^+$ and F$^-$ or Cs$^+$ and Cl$^-$ in Table 4.7; radii in Table 4.6).

Now it is time to consider deviations from the simple electrostatic pattern. Table 4.9 compares hydration enthalpies for some pairs of cations of identical charge and approximately equal radii. Values for B-Group and transition metal cations are markedly more negative than for A-Group ions of similar radii. Such differences are often rationalised, as indicated in Table 4.9, in terms of the Hard and Soft Acids and Bases, HSAB, approach (see section 6.2), with the extra hydration enthalpy ascribed to polarisation or covalent interaction contributions. For transition metal cations, the variation of crystal field stabilisation energies with d-electron configuration is reflected in ion hydration enthalpies (Table 4.10). For anions such as F$^-$, there is a

Table 4.9 — Cation hydration enthalpies for equivalent 'hard' and 'soft' cations

	'Hard'			'Soft'	
Cation	Radius (Å)	ΔH_{hydr} (kJ mol^{-1})	Cation	Radius (Å)	ΔH_{hydr} (kJ mol^{-1})
K^+	1.33	−321	Ag^+	1.29	−475
Mg^{2+}	0.65	−1922	Cu^{2+}	0.69	−2100
Ca^{2+}	0.99	−1592	Cd^{2+}	0.97	−1806
Sr^{2+}	1.13	−1445	Hg^{2+}	1.10	−1823

Table 4.10 — The reflection of Crystal Field stabilisation on ion hydration enthalpies for the first row d-block 2+ cations

Cation	Cr^{2+} d^4	Mn^{2+} d^5	Fe^{2+} d^6	Co^{2+} d^7	Ni^{2+} d^8	Cu^{2+} d^9	Zn^{2+} d^{10}
CFSE	$6Dq$	0	$4Dq$	$8Dq$	$12Dq$	$6Dq$	0
ΔH_{hydr}(kJ mol^{-1})	−1850	−1845	−1920	−2054	−2106	−2100	−2044

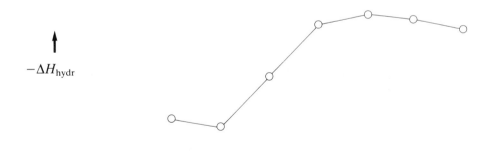

possibility of increased ion–solvent interaction through hydrogen-bonding in protic solvents such as water.

Table 4.11 shows how ion solvation enthalpies depend on the nature of the solvent. Such information is often presented in the form of enthalpies of transfer; the

Table 4.11 — Ion solvation enthalpies (kJ mol^{-1})

	Water	liq. NH$_3$	Methanol	Acetonitrile	Dimethyl sulphoxide	HMPA
Li$^+$	-515	-556	-531			
K$^+$	-321	-351	-351	-347	-368	-385
Cs$^+$	-263	-293				
Ag$^+$	-475	-577	-510	-536	-544	
Ba^{2+}	-1304	-1405	-1389	-1326	-1406	-1460
Cl$^-$	-369		-361	-349	-350	

enthalpy of transfer is simply the difference between the enthalpies of solvation of an ion in two solvents. Generally one of these solvents is water, to provide a common basis for comparisons. Formally the process of transfer involves taking the ion out of one solvent into the gas phase, then placing it in the second solvent (Fig. 4.8). Table

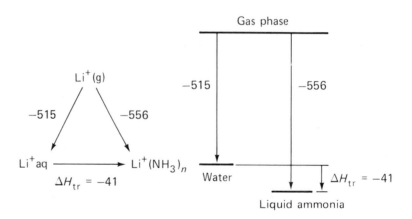

Fig. 4.8 — Enthalpy of transfer of Li$^+$ from water to liquid ammonia, and its relation to enthalpies of solvation.

4.12 includes a selection of ion transfer enthalpies, in all cases from water. Tables 4.11 and 4.12 show that solvation enthalpies for a cation such as K$^+$ or Ba^{2+} tend to be of the same order of magnitude, though strong donor solvents such as DMSO or HMPA do give markedly more favourable solvation enthalpies. However, for Ag$^+$ solvation by DMSO, MeCN, and ammonia is particularly favourable. Water solvates chloride effectively, while non-aqueous solvents generally are less effective;

Table 4.12 — Enthalpies of transfer (kJ mol^{-1}) of ions from water into various non-aqueous solvents

	Liq.NH$_3$	Methanol	Acetonitrile	Dimethyl sulphoxide	HMPA
K$^+$	-30	-30	-26	-47	-64
Ag$^+$	-102	-35	-61	-69	
Ba^{2+}	-101	-85	-22	-102	-156
Cl$^-$		$+8$	$+20$	$+19$	

this situation contrasts with the negative enthalpies of transfer for metal cations from water into non-aqueous media shown in Table 4.12.

The strengths of ion–solvent interactions are important in determining solubilities of salts, relative strengths of these interactions in determining differences in solubility of a given salt in various solvents. Thus, for example, potassium chloride is much less soluble in alcohols and in dimethyl sulphoxide than in water. This can be ascribed to much less favourable solvation of chloride in these non-aqueous solvents—in the case of dimethyl sulphoxide this effect dominates over the somewhat more favourable solvation of K$^+$ in dimethyl sulphoxide than in water. Potassium iodide and sodium nitrate have similar solubilities in water and in liquid ammonia, since significantly better solvation of K$^+$ and of Na$^+$ by the ammonia is offset by comparably poorer solvation of the I$^-$ and NO$_3^-$ by the ammonia. But silver iodide is very much more soluble—more than 10^8 times—in liquid ammonia than in water, because ammonia interacts much more strongly than water with the Ag$^+$ cation. Ions such as AsPh$_4^+$ and BPh$_4^-$ are fairly well solvated by many organic solvents, so [AsPh$_4$][BPh$_4$] is freely soluble in such solvents. But these large, singly charged, hydrophobic ions are so feebly solvated by water that [AsPh$_4$][BPh$_4$] is essentially insoluble in water despite its very small lattice energy.

Although enthalpies of solvation provide perhaps the simplest indicator of the strength of ion–solvent interactions, data on other thermodynamic functions are available and informative. Particularly for water and other structured, polar, hydrogen-bonding solvents, the entropy change associated with hydration (solvation) of an ion is an important quantity. Table 4.13 includes a selection of hydration entropies for simple and complex ions, and shows the role played by charge and radius. Large ions of low charge, such as Cs$^+$, I$^-$, ClO$_4^-$, NR$_4^+$, or AuBr$_4^-$, have large positive partial molal entropies. Their introduction into solvent water results in an increase in entropy or freedom, as they act as breakers of water structure. On the other hand, small ions, especially of charge 2$\pm$ or above, have negative partial molal entropies. Such ions are heavily hydrated. Here the transfer of water molecules from bulk solvent to the ion solvation shell, with consequent increase in ordering and loss of freedom, results in a nett decrease in entropy. This electrostriction of solvent under the influence of the electric field of the ion also results in a volume decrease (Fig. 4.9). Table 4.14 shows that partial molal volumes for ions in aqueous solution

Table 4.13 — Standard partial molal entropies ($J\,K^{-1}\,mol^{-1}$) for hydrated ions in aqueous solution (relative to zero for the proton)

Li^+	+11							
Na^+	+59	Mg^{2+}	−138	Al^{3+}	−322			
K^+	+101	Ca^{2+}	−53	Sc^{3+}	−255			
Rb^+	+120	Sr^{2+}	−33	Y^{3+}	−259			
Cs^+	+133	Ba^{2+}	+10	La^{3+}	−218			
		Zn^{2+}	−110	Ga^{3+}	−331			
Ag^+	+73	Cd^{2+}	−76	In^{3+}	−264			
Tl^+	+126	Hg^{2+}	−36					
		Ni^{2+}	−131	Fe^{3+}	−300			
				Ac^{3+}	−181	Th^{4+}	−423	
F^-	−10	OH^-	−11	SO_4^{2-}	+17			
Cl^-	+55	CN^-	−49	CrO_4^{2-}	+50			
Br^-	+80	NO_3^-	+125	$Cr_2O_7^{2-}$	+262			
I^-	+109	ClO_4^-	+182					
NH_4^+	+97	$Fe(CN)_6^{4-}$	+95					
NMe_4^+	+210	$Fe(CN)_6^{3-}$	+270					
NEt_4^+	+283					$AuBr_4^-$	+336	

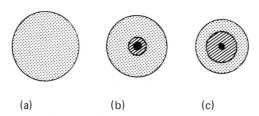

(a)   (b)   (c)

Fig. 4.9 — Illustration of the overall volume decrease when into a given volume of bulk solvent water (a) is introduced an ion of moderate (b) or large (c) charge. The darker hatching indicates electrostricted water molecules in the hydration shells of the ions.

show a similar pattern to partial molal entropies (Table 4.13).† Small and medium sized ions of charge 2± or more have negative partial molal hydration volumes dominated by electrostriction, but for large ions, particularly of charge only 1±, the sheer size of the ion dominates over the relatively small electrostriction contribution.

† The standard partial molal entropies and volumes in Tables 4.13 and 4.14 are based on the assumption of zero for the hydrated proton, an assumption which is believed to be close to reality in both cases. Absolute values for the partial molal entropy and volume of the hydrated proton are generally thought to be close to $-20\,J\,K^{-1}\,mol^{-1}$ and $-5\,cm^3\,mol^{-1}$ respectively.

Table 4.14 — Partial molal hydration volumes $(cm^3 mol^{-1})$ for hydrated ions (relative to zero for the proton)

Li^+	-0.9						
Na^+	-1.2	Mg^{2+}	-21.2	Al^{3+}	-42.2		
K^+	$+9.0$	Ca^{2+}	-17.9				
Rb^+	$+14.1$	Sr^{2+}	-18.2				
Cs^+	$+21.3$	Ba^{2+}	-12.5	La^{3+}	-39.1		
		Zn^{2+}	-21.6				
Ag^+	-0.7	Cd^{2+}	-20.0				
Tl^+	$+10.6$	Hg^{2+}	-19.3				
		Ni^{2+}	-24.0	Fe^{3+}	-44		
						Th^{4+}	-54
F^-	-1.1	OH^-	-4.0	SO_4^{2-}	$+14.0$		
Cl^-	$+17.8$	NCS^-	$+35.7$				
Br^-	$+24.7$	NO_3^-	$+29.0$				
I^-	$+36.2$	ClO_4^-	$+44.1$				
$[Co(NH_3)_6]^{3+}$	$+73$	$[Cr(ox)_3]^{3-}$	$+122$				
$[Co(NH_3)_5Cl]^{2+}$	$+94$	$[Fe(CN)_6]^{3-}$	$+121$				
		$[Fe(CN)_6]^{4-}$	$+74$				

In view of the similarities described above, it is hardly surprising to find that partial molal hydration entropies and volumes correlate quite closely. This is shown, for the case of cations in aqueous solution, in Fig. 4.10. Semi-empirical correlations are now being developed, in which, for example, partial molal hydration volumes for ions can be expressed as a function of their charge, radius, and hydration number. The application of these ideas to hydrated lanthanide(III) cations lent support to the hypothesis of a change in hydration number for these ions somewhere around the middle of the series.

Thus a fairly consistent picture of strengths of ion–solvent interactions can be built up from thermodynamic data which, on the whole, is compatible with the indications from ultraviolet–visible and infrared–Raman spectra (sections 4.1 and 4.2).

In recent years, understanding of fundamental thermodynamic aspects of ion solvation has been furthered by the determination of enthalpies and entropies for successive additions of water molecules to ions in the gas phase. Enthalpies for the typical cases of Li^+ and Cl^- are set out in Table 4.15. In each case the stepwise enthalpies decrease in magnitude as the number of water molecules attached to the ion increases, but this decrease is more gradual for Cl^- (and for other anions) than for Li^+ (and for other cations). By the time six water molecules have been attached

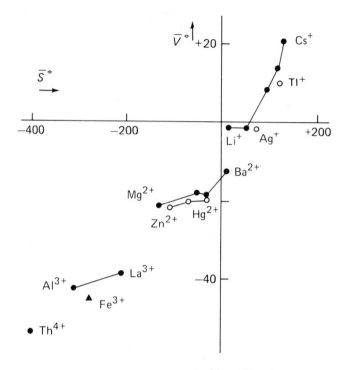

Fig. 4.10 — Relation between partial molar volumes $(\overline{V}^{\ominus}/cm^3\,mol^{-1})$ and entropies $(\overline{S}^{\ominus}/ J\,K^{-1}\,mol^{-1})$ for aqua-metal ions.

Table 4.15 — Enthalpy changes associated with the addition of water molecules to the Li^+ and Cl^- in the gas phase

	$\Delta H(kJ\,mol^{-1})$			
	Li^+		Cl^-	
	Stepwise	Cumulative	Stepwise	Cumulative
Addition of first	−142		−55	
second	−108	−250	−53	−108
third	−88	−338	−49	−157
fourth	−67	−405	−46	−203
fifth	−58	−463		
sixth	−50	−513		
water molecule				
Cf. single ion hydration enthalpy:		−515		−369

to Li^+, the overall enthalpy change is almost equal to the total enthalpy of hydration, i.e. the enthalpy of transfer of Li^+ from the gas phase into aqueous solution (see the discussion above). For chloride, the attachment of four water molecules is accompanied by an enthalpy change only just over half of that for transfer of Cl^- from the gas phase into aqueous solution.

5

Acid–base behaviour; hydrolysis and polymerisation

5.1 pK VALUES FOR AQUA-METAL IONS

Perhaps the simplest reaction of an aqua-cation in solution is the loss of a proton to give a hydroxo-aqua-species:

$$[M(OH_2)_x]^{n+} \rightleftharpoons [M(OH_2)_{x-1}(OH)]^{(n-1)+} + H^+ aq$$

A familiar example of this is provided by aluminium(III):

$$[Al(OH_2)_6]^{3+} \rightleftharpoons [Al(OH_2)_5(OH)]^{2+} + H^+ aq$$

Such behaviour leads on into the much more complicated area of polymerisation, for many hydroxo-aquacations showed marked tendencies to form hydroxo-, and oxo-, bridged species,† e.g.

$$2[Al(OH)]^{2+}aq \rightleftharpoons \left[Al\begin{smallmatrix} OH \\ OH \end{smallmatrix}Al \right]^{4+} aq.$$

or

† Polynuclear aqua-metal cations containing metal–metal bonds rather than hydroxo- or oxo- bridges are rare. The mercurous ion, $Hg_2^{2+}aq$, is the most familiar example. The rhodium(II) and molybdenum(II) species $Rh_2^{4+}aq$ and $Mo_2^{4+}aq$ are also stable entities in aqueous solution, but cations such as Cd_2^{2+} have only been detected in exotic solvents such as molten $NaCl$–$AlCl_3$.

The above equilibria have been set out for aqueous media, but proton loss and polymerisation are reactions which can also occur in protic non-aqueous solvents such as liquid ammonia. In practice very little is known about this aspect of non-aqueous solution chemistry, so this chapter is in fact confined to hydrolysis and polymerisation in aqueous solution.

At the simplest level, the coordination of a metal ion to a water molecule will, by electrostatics, make proton loss easier (Fig. 5.1). The greater the positive charge on

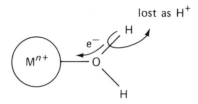

Fig. 5.1 — Acidity of a hydrated cation.

the ion, the easier it should be for the proton to dissociate from an attached water molecule.

The usual definitions of the equilibrium constants characterising such equilibria are given in the following equations:†

$$M^{n+} + OH^- = M(OH)^{(n-1)+} \qquad\qquad K_1 = \frac{[M(OH)^{(n-1)+}]}{[M^{n+}][OH^-]}$$

$$M^{n+} + H_2O = M(OH)^{(n-1)+} + H_3O^+ \quad {}^*K_1 = \frac{[M(OH)^{(n-1)+}][H_3O^+]}{[M^{n+}][H_2O]}$$

Table 5.1 includes values of hydrolysis equilibrium constants for a variety of metal cations. These values are at 298.2 K, and have been estimated for conditions of zero ionic strength by extrapolation. There are difficulties involved in measuring these equilibrium constants for many aqua-metal ions, and doubts as to the accuracy or precision of certain values are indicated by bracketing. Fuller details of methods, difficulties, uncertainties, ionic strength effects, and results can be found in the book by Baes and Mesmer cited at the end of this text.

The two most important features of the data included in Table 5.1 are the

† ${}^*K_1/K_1 = K_w$; $p{}^*K_1 = -\log {}^*K_1$; [H$_2$O] is taken as one. {Strictly equilibrium constants or quotients should be expressed in terms not of concentrations (activities) but of ratios of actual concentrations (activities) to concentrations (activities) in the standard state. In any normally dilute solution the activity of the water is very close to that of water in its standard state. This disposes of the seemingly arbitrary setting of [H$_2$O] = 1, but at the expense of having to specify the standard states assumed when giving the value of an equilibrium constant. The use of a one molar standard state for solute species results in the same numerical values as those obtained from using molar concentrations in the normally employed equations for equilibrium constants.}

Table 5.1 — Selected values of p^*K_1 for aqua-metal cations (at 298.2 K; molar scale; at zero ionic strength; values in brackets are very approximate)

sp-Block elements

Li^+	13.9	Mg^{2+}	11.4	Al^{3+}	5.0	Po^{4+}	(1?)
Na^+	14.7	Ca^{2+}	12.6	Sc^{3+}	(5)		
		Sr^{2+}	13.1	Y^{3+}	(8)		
Tl^+	13.3	Ba^{2+}	13.3	La^{3+}	(9)		
Ag^+	11.9						
		Zn^{2+}	9.5	Ga^{3+}	2.6		
		Cd^{2+}	7.9	Tl^{3+}	(1)		
		Sn^{2+}	1.9				

Transition metals

Cr^{2+}	9	Cr^{3+}	3.9
Fe^{2+}	(7)	Fe^{3+}	2.0
Ni^{2+}	10		
Pd^{2+}	1.4		

Lanthanides and actinides

Ce^{3+}	9.3	Ce^{4+}	(0)
Gd^{3+}	8.6		
Lu^{3+}	8		
		U^{4+}	1
Pu^{3+}	7.0	Pu^{4+}	1.5

enormous range of p^*K_1 values, and the marked dependence on ionic charge and radius. The first point is dramatically illustrated by comparing the values for the aquacations of the alkali metals with those for the 4+ cations. The alkali metal cations have minimal effect on the acidity of coordinated water — within experimental uncertainty their p^*K_1 values are equal to pK_w for water, 14. At the other extreme, the 4+ aquacations have acidities comparable with mineral acids — the pH of an ideal molar solution of a strong monobasic acid is zero. In general the 3+ cations make coordinated water as acidic as a weak organic acid such as acetic acid, while 2+ cations have only a very small effect in most cases. Such series of cations as $Mg^{2+} \rightarrow Ba^{2+}$ or $Al^{3+} \rightarrow La^{3+}$ show a much smaller but nonetheless real dependence on cation size, in the direction forecast by simple electrostatics.

Further inspection of Table 5.1 reveals that the application of the simple electrostatic rationalisation of aquacation p^*K_1 values is limited to the Group IA, IIA, IIIA and the *f*-block cations. Other aquacations are often more acidic, sometimes very much more acidic, than would be expected by comparison. Thus the large 1+ cations of silver and thallium have equilibrium constants for proton loss thousands and hundreds of times respectively greater than would be expected from the alkali metal cation values. Also transition metal 3+ aquacations and the Group IIIB aquacations Ga^{3+}, In^{3+}, and Tl^{3+} are considerably more acidic than their

Group IIIA analogues Al^{3+} to La^{3+}. Fig. 5.2 shows p^*K_1 values plotted against ionic radii for a number of 3+ aquacations. There is a great deal of scatter, indicating large deviations from a simple electrostatic pattern. Points for related ions have been connected by lines — thus the Group IIIA, Group IIIB, and first-row transition elements have been linked. The 3+ lanthanide ions fall on the line joining La^{3+} to Lu^{3+}; a parallel line links the actinide ions Np^{3+} and Pu^{3+}. Of all these groups, the Group IIIA sequence from Al^{3+} has the strongest claim to be considered as following a well-behaved electrostatically controlled trend. The exceptionally high acidity of the aqua-ions of cobalt(III), manganese(III), and thallium(III) is often ascribed to the highly oxidising properties of these metal ions — but the titanium(III) and bismuth(III) aquacations are also much more acidic than expected on simple electrostatic grounds.

There is an important chemical corollary to the high pK values of 3+ and, particularly, 4+, aquacations. That is that in order to have a solution of the aqua-ion which contains a negligible amount of hydroxo-aquacation, it is necessary to acidify the solution. For a cation such as Al^{3+}, one must add sufficient acid to reduce the pH to around 3, while for a 4+ cation with pK of around zero there will still be a small but significant concentration of hydroxo-species present even in a strong acid soluton.

So far only the loss of one proton from one coordinated water molecule in a given aquacation has been considered, to give a monohydroxoaquo species. However, it is possible to lose a second proton from a second water molecule in the primary hydration shell. This gives a bishydroxoaquo species, which may in turn lose further protons:

$$[M(OH_2)_x]^{n+} \overset{K_1}{\rightleftharpoons} [M(OH)(OH_2)_{x-1}]^{(n-1)+} + H^+$$

$$\overset{K_2}{\rightleftharpoons} [M(OH)_2(OH_2)_{x-2}]^{(n-2)+} + H^+ \text{ etc.}$$

Data (pK_n values) exist for the loss of a second and subsequent protons for a number of aqua-metal cations. In almost all cases there is a steady increase in p^*K_n as n increases. Sometimes the values of p^*K_1, p^*K_2, p^*K_3, etc. are well separated, so that at a given pH either one or two species predominate, but more often consecutive p^*K values are rather similar. In such circumstances three or more species may coexist in certain pH ranges. Thus, for example, Ga^{3+}aq has $p^*K_1 \simeq 2.9, p^*K_2 \simeq 4, p^*K_3 \simeq 4.5$; at a pH of around 3.5 there will be significant amounts of Ga^{3+}aq and of $Ga(OH)_2^+$aq in equilibrium with $Ga(OH)^{2+}$aq. In the exceptional case of Hg^{2+}, where $p^*K_1 = 3.7$ and p^*K_2 is actually *lower*, at 2.6, it has been calculated that at pH 3 the equilibrium composition is 59% Hg^{2+}aq, 12% $Hg(OH)^+$aq, and 29% $Hg(OH)_2$aq. For the many metal ions with sparingly soluble hydroxides, there may be difficulty in establishing a value for p^*K_n due to precipitation of uncharged $M(OH)_n$ (this may represent a special case of the formation of polynuclear hydroxo-bridged species, discussed in the next section).

At high pHs, metal ions may be coordinated entirely by hydroxide ligands, giving species of the type $[M(OH)_x]^{n-}$. Elements and oxidation states for which such

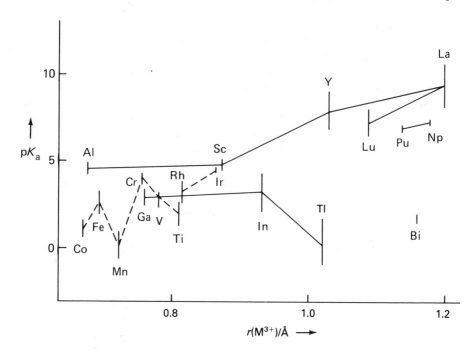

Fig. 5.2 — Relation between pK_a and ionic radius for metal cations.

species have been characterised or claimed are listed in Table 5.2, to give an idea of their widespread distribution around the Periodic Table. In several cases a given metal cation gives a range of stoichiometries. Thus for manganese(III), salts of $[Mn(OH)_5]^{2-}$, $[Mn(OH)_6]^{3-}$, and $[Mn(OH)_7]^{4-}$ have been isolated, while $[Cu(OH)_3]^-$, $[Cu(OH)_4]^{2-}$, and perhaps $[Cu(OH)_6]^{4-}$ exist in copper(II) solutions at high pHs. Silver(III) is only stable in aqueous media at high pHs, where it exists as the $[Ag(OH)_4]^-$ anion. The formation of these, and related species containing oxo-ligands, is the reason for the solubility of sparingly soluble hydroxides in strong alkalis such as sodium hydroxide solution.† Indeed the general method for preparing $[M(OH)_x]^{n-}$ anions is to reflux the appropriate hydroxide or hydrated oxide with very strong sodium or potassium hydroxide. Typically one would use a solution of 100 g sodium hydroxide in 100 g water — an aggressive reagent that boils at 143°C and requires the use of silver-lined apparatus. For high oxidation state complexes of this type, electrochemical preparations are possible; the $[Ag(OH)_4]^-$ anion mentioned above is generated by electrooxidation of a silver foil anode in sodium hydroxide solution.

Sparingly soluble metal hydroxides are, of course, usually soluble in aqueous acid. Solubility both in acidic and in alkaline aqueous media betokens amphoteric behaviour, some examples of which are included in Fig. 5.3 in the following section.

---

† Sodium hexahydroxoantimonate, $Na[Sb(OH)_6]$, is of interest as one of the very few sodium salts which are sparingly soluble in water.

Table 5.2 — Elements which form hydroxo-anions $[M(OH)_x]^{n-}$

	sp-Block			d-Block						f-Block	
M(II)	Be	Mg		Mn	Fe	Co	Ni	Cu	Zn		
	Sn	Pb							Cd		
M(III)	Al	Ga	In	Cr	Mn	Fe		Cu		Yb	Lu
	Sc		Bi				Rh	Ag			
M(IV)	Ge	Sn	Pb	Tc	Ru	Pd	Pt				
M(V)	Sb										

As an alternative to the stepwise loss of protons from successive coordinated water molecules, for ions of charge 4+ and higher it is possible to lose two protons from a single coordinated water molecule to give an oxo-aquacation, e.g.:

$$\text{"}[V(OH_2)_x]^{4+}\text{"} \rightarrow VO^{2+} aq + 2H^+$$

This process may be repeated, to give a dioxo-aquacation, as for vanadium(V) and a few actinides in oxidation state V or VI, e.g.:

$$\text{"}[U(OH_2)_x]^{6+}\text{"} \rightarrow UO_2^{2+} aq + 4H^+$$

It should be emphasised that this type of behaviour is relatively uncommon. Thus, for example, such species as Ce^{4+} and Th^{4+} give bis-hydroxo- rather than oxo-complexes on loss of two protons from one aqua-ion. Also the great majority of metals in oxidation states five and upwards do not have one or two oxo-ligands and several waters, they are generally simple oxoanions in which the metal is surrounded by several oxide ligands. Such are the familiar species permanganate, chromate, and vanadate, to name but three examples from many.

The choice between formation of hydroxo- or oxo- species depends to some extent on the different bonding properties of these ligands, and on electrostatic considerations. The O^{2-} ion has an advantage over OH^- in electrostatic stabilisation of high oxidation states, as in the vandium(IV) and uranium(VI) oxocations above. It is also easier for the O^{2-} ion to act as a π-donor ligand, as emphasised in classical depictions of, for example, permanganate as

and indeed UO_2^{2+} as $[O=U=O]^{2+}$. However, one has to be wary of simple explanations in this area. Thus the π-donor theory has to be applied with some subtlety to accommodate linear (trans) MO_2^{n+} units for the actinides, but cis-dioxo geometry in VO_2^+ and in $[ReO_2(CN)_4]^-$. The balance betwen O^{2-} and OH^- must be fairly fine in polynuclear species, as will become apparent in the following section.

5.2 POLYMERISATION

It is not always easy to measure the pK values for aquacations accurately. While in the cases of the weakest and strongest aquacation acids this is purely a technical problem, in several instances there is a more fundamental chemical difficulty. This is the readiness with which many hydroxo-aquacations polymerise to give polynuclear hydroxo-bridged cations, to be described in the following paragraphs. The possibility or likelihood of such polynuclear cation formation may make it necessary to restrict measurements to a small pH range to avoid interference from the often slow polymerisation reactions. Indeed it is likely that for, e.g., In^{3+}, polynuclear cation formation may interfere under all conditions, making accurate determination of p*K_1 simply for loss of one proton well-nigh impossible.

Polynuclear complex formation can be significant for the 2+ cations of the transition metals in approximately neutral solution. At pHs well below 7 these cations exist simply as M^{2+}aq, while in alkaline solution they precipitate as the sparingly soluble hydroxides. But at pHs a little below the onset of $M(OH)_2$ precipitation, there will be considerable concentrations of polynuclear hydroxo-cations. The precise pHs involved vary with the nature of the metal as well as its concentration — at pH 6 more than 5% of copper(II) is present as a dinuclear cation, whereas for nickel the pH has to be increased to 8 before about 1% of polynuclears are formed. A similar situation obtains for 3+ and 4+ cations, though of course the pH range for polynuclear formation and for hydroxide precipitation will be considerably lower than those for the 2+ cations. The acid limit will be determined by the p*K_1 for the aquacation, the alkaline limit by the solubility product for the hydroxide.

The proposed formulae for some of the hydroxo-aquapolynuclear cations derived from aluminium(III), bismuth(III), nickel(II), and lead (II) are set out in Fig. 5.3. This shows the structures of a few of these species. It also shows the relation of the polynuclear species to the hydroxo-aquacations and the parent aquacations. It includes a reminder that at high pHs there will be precipitation of hydroxides, which may even dissolve in strongly alkaline media to give oxoanions such as aluminates and plumbites.

The four elements selected for inclusion in Fig. 5.3 have a particularly rich polymerisation chemistry in aqueous media. However, the existence of hydroxo-bridged polynuclear species is established or suspected for the great majority of 3+ and 4+ ions and for many 2+ ions. Details relating to ions other than those included in Fig. 5.3 can be found in the book by Baes and Mesmer cited in the Further Reading section at the end of this book. This also outlines the quantitative treatment of polymerisation equilibria of this type, a topic too complicated for coverage here. As in some other contexts cited earlier, there is a surprising lack of detailed knowledge and understanding of some seemingly rather simple and very fundamental topics in this whole area of the solution chemistry of metal ions.

We have concentrated on relatively low oxidation states in this section so far. The formation of polynuclear species is also important for many metals in high oxidation states, though now the bridging is mainly or exclusively through oxide bridges. As a

Polymerisation

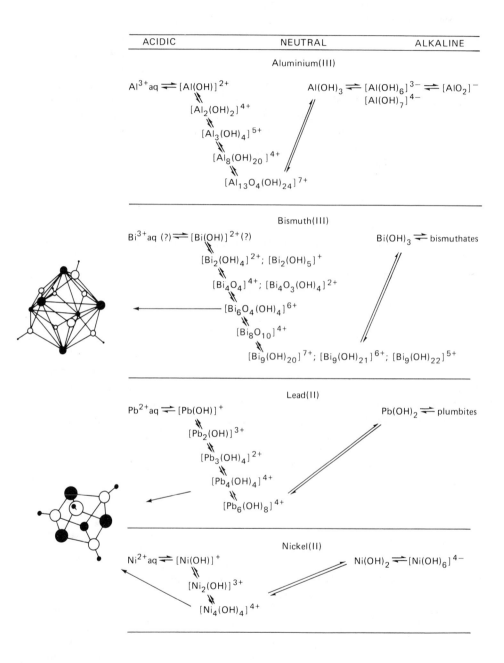

ACIDIC	NEUTRAL	ALKALINE

Aluminium(III)

$Al^{3+}aq \rightleftharpoons [Al(OH)]^{2+}$

$[Al_2(OH)_2]^{4+}$

$[Al_3(OH)_4]^{5+}$

$[Al_8(OH)_{20}]^{4+}$

$[Al_{13}O_4(OH)_{24}]^{7+}$

$Al(OH)_3 \rightleftharpoons [Al(OH)_6]^{3-} \rightleftharpoons [AlO_2]^-$

$[Al(OH)_7]^{4-}$

Bismuth(III)

$Bi^{3+}aq \; (?) \rightleftharpoons [Bi(OH)]^{2+} \; (?)$

$[Bi_2(OH)_4]^{2+}; \; [Bi_2(OH)_5]^+$

$[Bi_4O_4]^{4+}; \; [Bi_4O_3(OH)_4]^{2+}$

$[Bi_6O_4(OH)_4]^{6+}$

$[Bi_8O_{10}]^{4+}$

$[Bi_9(OH)_{20}]^{7+}; \; [Bi_9(OH)_{21}]^{6+}; \; [Bi_9(OH)_{22}]^{5+}$

$Bi(OH)_3 \rightleftharpoons bismuthates$

Lead(II)

$Pb^{2+}aq \rightleftharpoons [Pb(OH)]^+$

$[Pb_2(OH)]^{3+}$

$[Pb_3(OH)_4]^{2+}$

$[Pb_4(OH)_4]^{4+}$

$[Pb_6(OH)_8]^{4+}$

$Pb(OH)_2 \rightleftharpoons plumbites$

Nickel(II)

$Ni^{2+}aq \rightleftharpoons [Ni(OH)]^+$

$[Ni_2(OH)]^{3+}$

$[Ni_4(OH)_4]^{4+}$

$Ni(OH)_2 \rightleftharpoons [Ni(OH)_6]^{4-}$

Fig. 5.3 — Interrelations between aquacations, hydroxo-cations, polynuclear cations, hydroxides, hydroxo-anions, and metallates for selected elements.

result the polynuclear ions are generally negatively charged. One of the most familiar examples is that of chromium. In alkaline media chromium(VI) exists as chromate, $[CrO_4]^{2-}$, but in dilute acid the stable form is dichromate, $[Cr_2O_7]^{2-}$. At sufficiently low pHs the trinuclear anion $[Cr_3O_{10}]^{2-}$ can be formed, while treatment of $[Cr_3O_{10}]^{2-}$ with more chromium trioxide in concentrated nitric acid yields $[Cr_4O_{13}]^{2-}$, which can be isolated as its potassium salt. Molybdenum(VI), tungsten(VI), and vanadium(V) also exist in the form of the respective mononuclear oxoanions in alkaline solution, but as polynuclear anions in acidic media. The state of aggregation tends to be larger for these elements than for chromium. There is a decavanadium oxoanion, while for tungsten the following sequence of increasing aggregation has been suggested:

$$WO_4^{2-} \rightarrow W_4O_{14}^{4-} \rightarrow W_6O_{21}^{6-} \rightarrow W_{12}O_{41}^{10-}$$

Polyanions containing up to 48 tungsten atoms have been claimed, while two Mo_{36} species have recently been characterised. There is an enormous chemistry of iso- and of hetero- polytungstates and polymolybdates, in the solid state even more than in solution.

In contrast to the complicated chemistry outlined in the preceding paragraph, manganese(VII) and rhenium(VII) exist as the simple MnO_4^- and ReO_4^- anions over the whole pH range. There is a similar pattern amongst oxoanions of the *sp*-block elements, with a whole range of polynuclear phosphates, silicates, and borates, but with perchlorate totally resistant to polynuclear anion formation. The oxoanion chemistry of sulphur (Fig. 5.4) is particularly varied, since in addition to polynuclear species with oxo- and peroxo- bridges there is a series of polynuclear anions with short chains of sulphur atoms (cf. S_8 for elemental sulphur). A mixture of polythionato-anions can be generated by passing hydrogen sulphide through a saturated aqueous solution of sulphur dioxide at 0°C; there are also specific preparative methods for individual members of the series. The tetrathionate anion is particularly commonly encountered, since it is a product in the volumetric determination of iodine using thiosulphate:

$$I_2 + 2S_2O_3^{2-} \rightarrow 2I^- + S_4O_6^{2-}$$

Dithionate has proved useful as counterion in isolating a number of cationic complexes from aqueous solution (section 1.4).

The two topics of the preceding paragraph come together in the developing field of polynuclear cations with sulphide instead of oxide bridges. Thus there is a sequence from mononuclear thio-molybdenum species through such entities as $[Mo_3S_{13}]^{2-}$ to the limiting case of molybdenum disulphide, and a variety of clusters such as the $Fe_4S_4^{4+}$ unit (with the distorted cube structure of $[Ni_4(OH)_4]^{4+}$ and $[Pb_4(OH)_4]^{4+}$ (Fig. 5.3)) important in biochemical electron transfer, related species such as $Mo_4S_4^{6+}$ and $Mo_3S_4^{4+}$ (cf. $Mo_3O_4^{4+}$), and hybrid oxo- and thio-bridged species such as $Mo_3O_2S_2^{4+}$ aq.

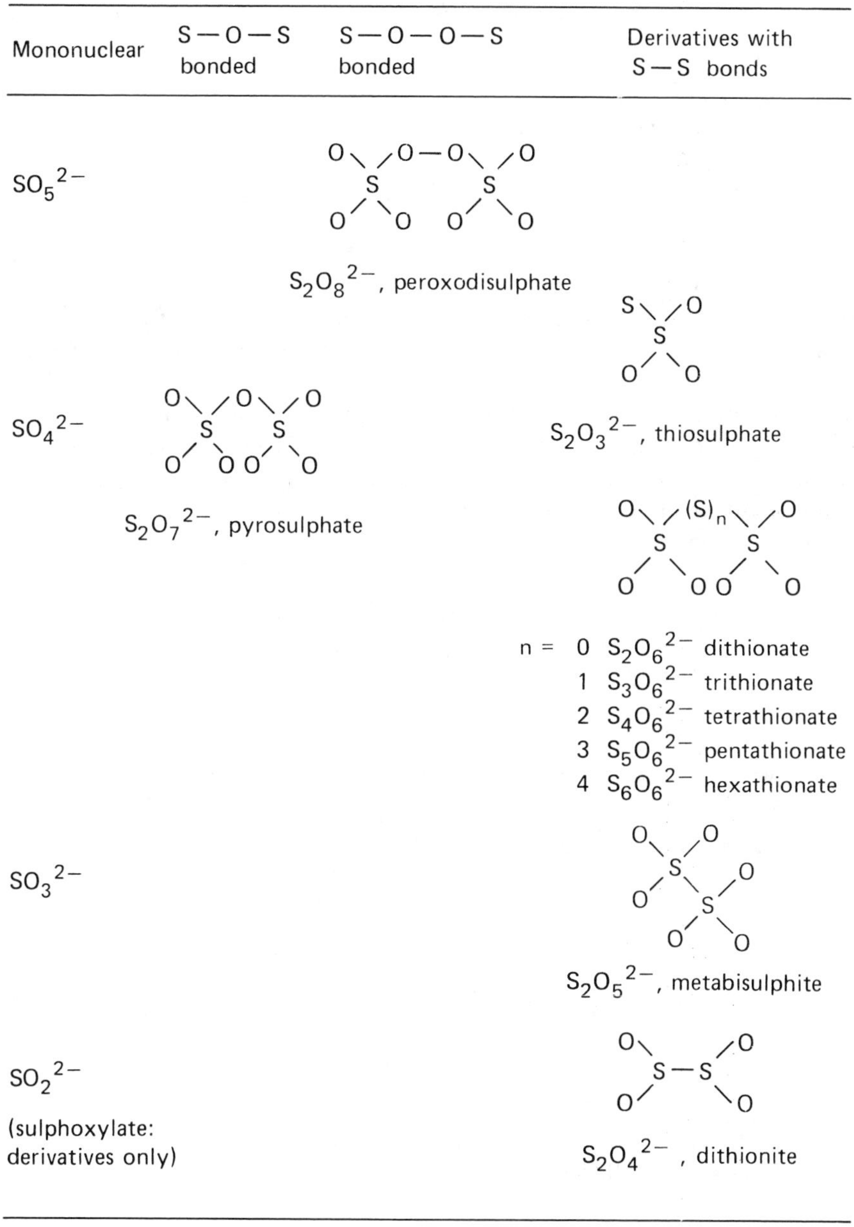

Fig. 5.4 — Oxoanions of sulphur.

5.3 ANIONS AND LIGANDS

Water molecules, and indeed protic ligands in general, when coordinated to metal ions, are prone to proton loss. Conversely, anionic ligands may be subjected to protonation. Simple anions such as chloride, perchlorate, nitrate, or sulphate are derived from strong acids, and thus will become protonated only at very low pHs.

But the majority of ligands are the conjugate bases of weak acids, and equilibria such as

$$CN^- + H^+ \rightleftharpoons HCN$$
$$CH_3CO_2^- + H^+ \rightleftharpoons CH_3CO_2H$$
$$NH_3 + H^+ \rightleftharpoons NH_4^+$$
$$py + H^+ \rightleftharpoons pyH^+$$

play an important role in their aqueous solution chemistry. If such ligands are added to an acidic aqueous solution containing potentially complexing metal ions, they may become protonated and thus unable to form the metal complex. Such behaviour is especially likely to be troublesome when trying to prepare complexes of metal ions with low p^*K_1s.

Some common series of complexes, for instance of salicaldehyde, of dimethylglyoxime, of acetylacetone (pentane-2,4-dione), or of cyclopentadiene, are best prepared in basic solution, since in all these cases it is the anion rather than the neutral reagent which forms stable complexes (Fig. 5.5).

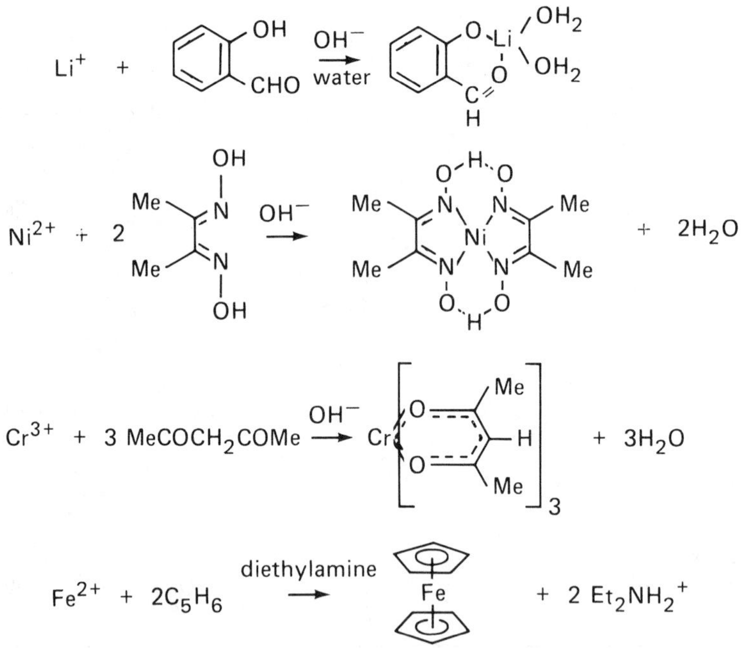

Fig. 5.5 — Generation of complexes of anionic ligands from their conjugate acids.

5.4 COMPLEXES

This chapter closes with two generalisations of the picture developed in section 5.1 with particular relevance to aquacations. The first is the extension from an aquacation $[M(OH_2)_x]^{n+}$ to the more general case of ternary species $[M(OH_2)_{x-y}L_y]^{n\pm}$. The important point to make in this connection is that the ligands L will have a significant effect on the acidity of the water ligands. In a few cases, for example when $L = CN^-$, the effect is dramatic. The orders of magnitude of these ligand effects are shown, for some cobalt(III) and chromium(III) complexes, in Table 5.3.

Table 5.3 — Acidity constants for aqua-ligand complexes

	Cr^{III}	Co^{III}
$[M(CN)_5(OH_2)]^{2-}$	ca. 9	9.7
$[M(NH_3)_5(OH_2)]^{3+}$	5.1	6.2
cis-$[M(NH_3)_4(OH_2)_2]^{3+}$	5.3	5.9
cis-$[M(en)_2(OH_2)_2]^{3+}$	4.8	6.1
trans-$[M(en)_2(OH_2)_2]^{3+}$	4.2	4.6
cis-$[M(phen)_2(OH_2)_2]^{3+}$		4.7
cis-$[Cr(ox)_2(OH_2)_2]^-$	5.6	
$[MCl(OH_2)_5]^{2+}$	5.2	
cf. $[M(OH_2)_6]^{3+}$	3.9	ca. 1

The acidity of protic ligands other than water will also be affected by coordination to a metal ion, though the effect may not always be as readily apparent as it is in the case of water. Thus the p^*K_1 values for the $[Co(NH_3)_6]^{3+}$ and $[Rh(NH_3)_6]^{3+}$ cations are > 14. There is only an extremely low, indeed not directly detectable, concentration of the conjugate base species $[M(NH_2)(NH_3)_5]^{2+}$ (though such conjugate species play a key role in base hydrolysis of this type of complex — see section 11.2). Only for the ammines of highly charged platinum(IV) and strongly oxidising gold(III) are there significant quantities of conjugate base species in basic aqueous solution (p^*K_1 for $[Pt(NH_3)_6]^{4+} \simeq 8$; for $[Au(NH_3)_4]^{3+} = 7.5$).

The converse to the above behaviour is that a ligand such as cyanide is much harder to protonate when coordinated to a metal ion in a complex than when free in solution. It is much more difficult to generate $[Fe(CN)_5(CNH)]^{3-}$ than HCN. In similar vein, the carbonato-complex $[Co(en)_2(CO_3)]^+$ exists as such in aqueous solutions wherein uncomplexed carbonate would exist as CO_3H^-.

6

Stability constants

6.1 DEFINITIONS

One of the most important reactions of solvated metal ions is the replacement of one or more solvent molecules by ligand molecules or ions to give complexes (Fig. 6.1).

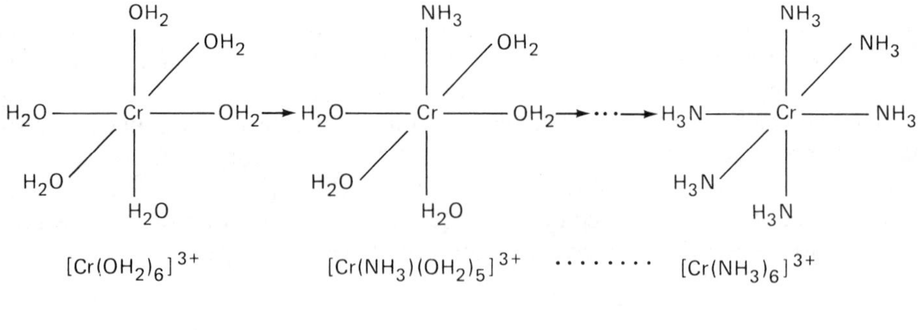

Fig. 6.1 — The stepwise conversion of hexaaquachromium(III) into hexaamminechromium(III).

Strengths of metal ion–ligand interactions are related to strengths of metal ion–solvent (almost always ion–water) interactions by the use of stability (formation) constants. These are defined in relation to the appropriate formation equilibria involved as shown in the following equations:

Stepwise

$$M + L = ML \qquad K_1 = \frac{[ML]}{[M][L]}$$

$$ML + L = ML_2 \qquad K_2 = \frac{[ML_2]}{[ML][L]}$$

$$ML_2 + L = ML_3 \qquad K_3 = \frac{[ML_3]}{[ML_2][L]}$$

$$\downarrow \qquad \downarrow \qquad\qquad \downarrow \qquad \downarrow$$

$$ML_n + L = ML_n \qquad K_n = \frac{[ML_n]}{[ML_{n-1}][L]}$$

Overall

$$M + L = ML \qquad \beta_1 = \frac{[ML]}{[M][L]}$$

$$M + 2L = ML_2 \qquad \beta_2 = \frac{[ML_2]}{[M][L]^2}$$

$$M + 3L = ML_3 \qquad \beta_3 = \frac{[ML_3]}{[M][L]^3}$$

$$\downarrow \qquad \downarrow \qquad\qquad \downarrow \qquad \downarrow$$

$$M + nL = ML_n \qquad \beta_n = \frac{[ML_n]}{[M][L]^n}$$

The nth overall stability (formation) constant, β_n, is simply the product of the first n stepwise constants, $\beta_n = K_1 K_2 \ldots K_n$. Ideally these equilibrium constants should be defined in terms of activities, but in practice stability constants for complexes in solution are expressed in terms of the experimental variable concentration.†

6.2 TRENDS

The normal trend for stepwise stability constants K_n is to decrease as n increases, as shown in Table 6.1, where the examples cover sp-block and d-block metal ions and anionic and neutral (monodentate and bidentate) ligands. Sometimes this trend is determined almost entirely by the entropy variation, as for copper(II)–ammonia (Table 6.2), sometimes by both enthalpy and entropy, as for aluminium(III)–fluoride (Table 6.2). When K_n is observed not to decrease smoothly with n, then there is either a change in coordination number (octahedral/tetrahedral or linear) or in magnetic moment (high spin/low spin) of the metal ion. Three such abnormal sequences are illustrated and compared with normal sequences for closely related

† Conventionally water is ignored both in the formation equilibria and the defined equilibrium constants (see the footnote in section 5.1, page 63)

Table 6.1 — Dependence of stepwise stability constants ($\log_{10}K_n$; molar scale) on the number of ligands (n) for typical systems (at 298.2 K)

Metal ion/ligand	Al^{3+}/F^-	Co^{2+}/NH_3	Cr^{3+}/NCS^-	Pb^{2+}/Cl^-	Fe^{2+}/en
$\log_{10}K_1$	6.1	2.1	3.0	1.2	4.3
$\log_{10}K_2$	5.0	1.6	1.7	0.5	3.3
$\log_{10}K_3$	3.9	1.1	1.0	−0.3	2.0
$\log_{10}K_4$	2.7	0.7	0.3	−1.3	
$\log_{10}K_5$	1.6	0.2	−0.7		
$\log_{10}K_6$	0.5	−0.6	−1.3		

en = $H_2NCH_2CH_2NH_2$.

Table 6.2 — Enthalpy and entropy contributions to stability constant trends

Copper(II)–ammonia

n	1	2	3	4
$\log_{10}K_n$	4.1	3.5	2.9	2.1
$\Delta G^\ominus$/kJ mol^{-1}	−23	−20	−15	−11
$\Delta H^\ominus$/kJ mol^{-1}	−23	−23	−23	−22
$T\Delta S^\ominus$/kJ mol^{-1}	0	−3	−7	−10

Aluminium(III)–fluoride

n	1	2	3	4	5	6
$\log_{10}K_n$	6.1	5.0	3.9	2.7	1.6	0.5
$\Delta G^\ominus$/kJ mol^{-1}	−35	−29	−21	−15	−7	−2
$\Delta H^\ominus$/kJ mol^{-1}	+5	+3	+1	+1	−1	−6
$T\Delta S^\ominus$/kJ mol^{-1}	+40	+32	+22	+16	+6	−4

systems in Fig. 6.2. $K_2 \simeq K_1$ in the nickel(II)–dimethylglyoxime system, since the interligand intramolecular hydrogen bonding which stabilises [Ni(dmgH)$_2$] (see Fig. 5.5 on page 72) is not possible in the mono-complex [Ni(dmgH)]$^+$.

The magnitudes of K_n values show a complicated dependence on the nature of the metal ion and on the nature of the ligand. The main part of the metal–ligand bond strength derives from the σ-bonding, whose extent depends on the donor properties of the ligand and on the acceptor properties of the metal ion. As described in section 4.3, there is no simple explanation for the observed strengths of metal ion–water interactions. As stability constants represent the *differences* between metal ion–water and metal ion–ligand interactions, it follows that there will be no simple universal rationalisation. Moreover there is the added complication of significant π-bonding between certain combinations of ligands and of metal ions. Despite all these complications there are some general guidelines which may be used to correlate

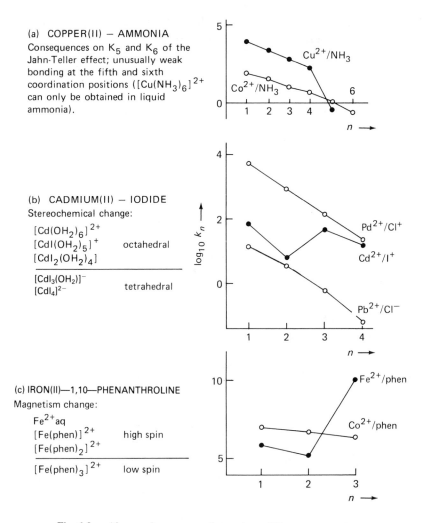

(a) COPPER(II) — AMMONIA
Consequences on K_5 and K_6 of the Jahn-Teller effect; unusually weak bonding at the fifth and sixth coordination positions ($[Cu(NH_3)_6]^{2+}$ can only be obtained in liquid ammonia).

(b) CADMIUM(II) — IODIDE
Stereochemical change:

$[Cd(OH_2)_6]^{2+}$
$[CdI(OH_2)_5]^+$ octahedral
$[CdI_2(OH_2)_4]$
_____
$[CdI_3(OH_2)]^-$
$[CdI_4]^{2-}$ tetrahedral

(c) IRON(II)—1,10—PHENANTHROLINE
Magnetism change:

Fe^{2+}aq
$[Fe(phen)]^{2+}$
$[Fe(phen)_2]^{2+}$ high spin
_____
$[Fe(phen)_3]^{2+}$ low spin

Fig. 6.2 — Abnormal sequences of stepwise stability constants, K_n.

stabilities. The most used is that of the Hard and Soft Acids and Bases (HSAB) principle, outlined in the following paragraph.

The HSAB principle, now applied in many and varied areas of chemistry, was first developed to provide a qualitative framework for rationalisation of the complicated field of stabilities of complexes in aqueous solution. It grew out of Chatt and Ahrland's division of metal ions and of ligands into Class 'a' and Class 'b' categories on the basis of stability constant trends. Class 'a' metal ions were those whose stability constants decreased in the orders:

$F^- \gg Cl^- > Br^- > I^-$

O ligands $\gg$ S ligands

N ligands $\gg$ P ligands.

Class 'b' metal ions were those exhibiting the opposite trends. This Class 'a' and 'b' character is shown in Fig. 6.3, and metal ions assigned to their respective groups in

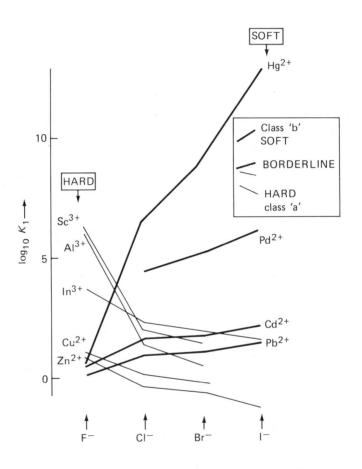

Fig. 6.3 — Stability constant trends and their relation to the the Class 'a'/Class 'b' and Hard and Soft Acids and Bases classifications.

Table 6.3. The Class 'a' and Class 'b' classification of ligands was complementary, with F^- and oxygen and nitrogen donor ligands labelled as Class 'a', the other halides and sulphur and phosphorus donor ligands as Class 'b'. The guiding principle was then that like ion–ligand pairs ('a' + 'a' or 'b' + 'b') gave stable complexes, unlike combinations ('a' + 'b') gave complexes of lower stability. Pear-

Table 6.3 — Classifications of metal ions

	$\log_{10}K_1$ (F$^-$) $-$ $\log_{10}K_1$ (Cl$^-$)		HSAB	Approximate geochemical equivalents
CLASS 'a'	> 5 4 to 5 1 to 4	Be^{2+} Sc^{3+} Zr^{4+} Th^{4+} U^{4+} Sn^{2+} Cr^{3+} Fe^{3+} Y^{3+} ^{a} Mg^{2+} La^{3+} Ce^{3+} Ac^{3+}	HARD	lithophilic (silicate phases)
	0 to 1	Pb^{2+} Fe^{2+} $\rightarrow$ Cu^{2+}		
	0 0 to -1	Tl$^+$ Cd^{2+}	BORDER-LINE	siderophilic^b
CLASS 'b'	< -1	Ag$^+$ Hg^{2+} (Pd^{2+} Pt^{2+})^c	SOFT	chalcophilic (sulphide phases)

^{a}VO^{2+} and UO$_2^{2+}$ also come into this category.
^{b}Noble metals such as palladium and platinum come into this category rather than into the 'soft' chalcophilic category, since they occur in nature as native metals.
^{c}These cations should undoubtedly come into this category (see Fig. 6.3), though in practice the stability constants for their fluoride complexes are not known.

son popularised this approach as the HSAB principle; 'Hard' = Class 'a' and 'Soft' = Class 'b'. Fig. 6.3 and Table 6.3 indicate that there are varying degrees of hardness and softness, with Tl$^+$ and several transition metal 2 + cations on the borderline. Fig. 6.3 shows that Fe^{2+} and Sc^{3+} are evidently much harder than In^{3+}, Cu^{2+}, and Zn^{2+}, while Cd^{2+} and Pb^{2+} are marginally soft and Hg^{2+} very soft. These broad guidelines have proved very popular rationalisations, and of some use in a qualitative predictive way. Quantitative expressions with appropriate equations and parameters, and a real understanding of the various chemical factors and their interplay, are still in an early stage of development.

Semi-quantitative correlations of stability constants with various ligand and metal ion properties are often quite successful when they are restricted in scope to certain groups of ligands or metal ions. Thus correlations between stability constants and ligand basicities can often be demonstrated for limited sets of ligands, especially when π-bonding can be neglected or when it correlates with σ-bonding. Such a correlation, covering several powers of 10 in K_1 and in ligand pK_a, holds for copper(II) complexes of substituted salicaldehydes. If one examines the variation in K_1, and the associated enthalpy for complex formation, with metal ion for the first-row transition element 2 + or 3 + ions, then there is a clear dependence on crystal field stabilisation energies (Table 6.4 and Fig. 6.4). Comparison of Fig 6.4(b) with Fig 6.4(a) shows the expected dominance of formation enthalpies; the $T\Delta S$ term spans only 3 kJ mol^{-1} for this series of ethane-1,2-diamine complexes. Plots of the type shown in Fig. 6.3(a) are representative of the frequently encountered 'Irving–Williams order' of stability constants (see the Glossary for details). The exact form of plots of the type shown in Fig. 6.4 may be affected by non-octahedral coordination for Zn^{2+}, and by Jahn–Teller contributions for Cr^{2+} and Cu^{2+}. The crystal field effect on stability constants represents the *difference* between the crystal field effects of the ligand and of water. If Jahn–Teller distortions are different in a

Table 6.4 — Crystal Field effects on stability constants

Metal ion	V^{2+}	Cr^{2+}	Mn^{2+}	Fe^{2+}	Co^{2+}	Ni^{2+}	Cu^{2+}	Zn^{2+}
Configuration	d^3	d^4	d^5	d^6	d^7	d^8	d^9	d^{10}
CFSE	$12Dq$	$6Dq$	0	$4Dq$	$8Dq$	$12Dq$	$6Dq$	0
$\log_{10}K_1(\text{en})$	4.6	5.2	2.7	4.3	5.9	7.6	10.7	5.9
$\log_{10}K_1(\text{edta})$	12.7	12.8	12.3	14.2	16.1	18.5	18.8	16.5

en = ethane-1,2-diamine.
edta = ethane-1,2-diaminetetraacetate.

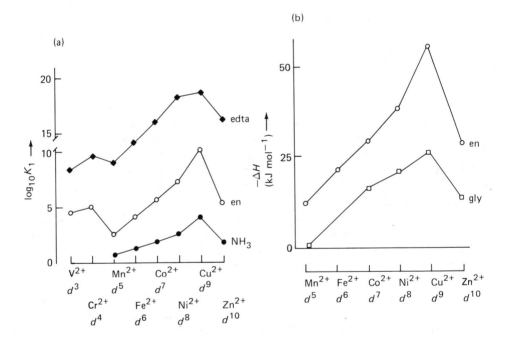

Fig. 6.4 — The Irving–Williams dependence of stability constants and formation enthalpies on
d electron configuration for high-spin complexes of first-row transition metal 2 + cations.

complex and in its parent aqua-ion, then the shape of Fig. 6.4 plot will be markedly
affected.

6.3 CHELATES

The presence of chelate rings confers an extra degree of stability on a complex, with
stability increasing as the number of chelate rings increases. Thus for a given metal
ion M^{n+} the stability constant for the tetradentate ligand LLLL is greater than the

overall stability constant for the bis-bidentate complex $M(LL)_2^{n+}$, in turn greater than the overall stability constant for the tetra-monodentate complex ML_4^{n+} (Fig. 6.5). This is known as the 'chelate effect'; the 'macrocyclic effect' is the further

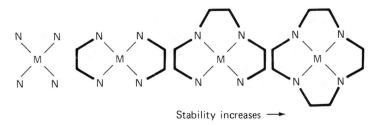

Stability increases ⟶

Fig. 6.5 — The chelate and macrocyclic effects.

stability conferred by a cyclic as opposed to a garland (linear) ligand of equal denticity and analogous structure (Fig. 6.5). In practice it is difficult or impossible to make *exact* comparisons, for neither ammonia nor methylamine are precisely equivalent to half of ethane-1,2-diamine. It is impossible to avoid complications such as the difference between primary and secondary nitrogen atoms, or the presence or absence of electron-releasing methyl groups. Analogous comments apply to 'equivalent' cyclic and linear polydenate ligands. Most textbooks emphasise the role of entropy in the chelate effect, illustrating with examples such as that given in Table 6.5, but the reader should beware of simple explanations of the chelate effect and of

Table 6.5 — The chelate effect

(a) Comparison of tetradentate, bis-bidentate, and tetrakis-monodentate complexes:

	$4\,NH_3$	(en)	(trien)
	$\log_{10}\beta_4$	$\log_{10}\beta_2$	$\log_{10}\beta_1$
Cd^{2+}	~7.5	10.1	10.8
Co^{2+}	4.7	10.7	11.2
Cu^{2+}	13.1	19.9	20.5

(b) Enthalpy and entropy contributions:

$$\text{For } Cd(NH_3)_2^{2+} + en \rightarrow Cd(en)^{2+} + 2\,NH_3$$
$$\Delta G^{\ominus} = -5.0\ \text{kJ mol}^{-1}$$
$$\Delta H^{\ominus} = +0.4\ \text{kJ mol}^{-1}$$
$$T\Delta S^{\ominus} = +5.4\ \text{kJ mol}^{-1}$$

the macrocyclic effect. Not only are there the difficulties over choosing exactly equivalent sets of ligands mentioned above, but there are also thermodynamic niceties such as the dependence of the magnitude of the chelate effect on the standard state chosen. Moreover enthalpy changes also often play a significant role in the macrocyclic effect.

There are no similar complications over the next general feature of stability constants, that five- and six-membered chelate rings are usually the most stable, with stability then decreasing as ring size increases (though see next but one paragraph for some exceptions to this general rule). Sequences of stability constants for two homologous series of ligands are shown in Table 6.6 to illustrate the general trend.

Table 6.6 — Dependence of stability constants on chelate ring size

Ligands	n	Chelate ring size	$\log_{10}K_1$		
			Mn^{2+}	Ca^{2+}	Zn^{2+}
oxalate homologues	0 (oxalate)	5	2.93		5.0
	1 (malonate)	6	2.30		3.7
$[O_2C(CH_2)_nCO_2]^{2-}$	2 (succinate)	7	1.26		2.5
	3 (glutarate)	8	1.13		2.0
	5 (pimelate)	10	1.08		
	7 (azelate)	12	1.03		
edta homologues $\begin{bmatrix} O_2CCH_2 \diagdown \quad \diagup CH_2CO_2 \\ \quad N(CH_2)_nN \\ O_2CCH_2 \diagup \quad \diagdown CH_2CO_2 \end{bmatrix}^{4-}$	2	5		10.70	
	3	6		7.12	
	4	7		5.05	
	5	8		4.40	

Again entropy plays an important role, qualitatively visualisable in terms of greater loss of freedom on chelate ring closure as ligand length increases. As the size of the potential chelate ring increases, then in practice long chain bidentate ligands often prefer to act as monodentate ligands or even as bridging ligands in binuclear species. There are a number of illustrations of this in the chemistry of long chain diamine complexes of cobalt(III).

Four-membered chelate rings are generally much less stable than five-membered rings, particularly for *sp-* and *d*-block cations. However, there are a number of complexes known which contain oxoanions such as sulphate, carbonate, or nitrate acting as bidentate ligands. Thus cobalt(III) forms $[Co(CO_3)(NH_3)_4]^+$ as well as $[Co(CO_3)(NH_3)_5]^+$, containing bidentate and monodentate carbonate respectively. Bidentate coordination of such oxoanions is more common in complexes of *f*-block metal ions. There are a number of complexes of this type where the metal has a high coordination number, often twelve. Thorium(IV) in the $[Th(NO_3)_6]^{2-}$ anion and lanthanum(III) in $La_2(SO_4)_3 9H_2O$ are both coordinated by twelve oxygen atoms from six bidentate nitrate or sulphate ligands. The well-known analytical reagent ammonium cerium(IV) nitrate is in fact the ammonium salt of the anionic complex $[Ce(NO_3)_6]^{2-}$, which has six bidentate nitrate ligands bonded to the cerium.

There is one small area where normal chelating ligands may not confer extra stability. A few cations, including silver(I) and mercury(II), favour a coordination number of two, with a linear arrangement of ligands. In such a situation it is only possible for a bidentate ligand to form a chelate ring if the two donor atoms are separated by perhaps six or eight other atoms. Only such a long chain permits coordination at two sites on opposite sides of a metal ion, and, as mentioned above, the formation of such large chelate rings is rather disfavoured. A ligand such as ethane-1,2-diamine cannot span *trans* positions, and acts as a monodentate ligand towards Ag^+ (log β_2 for $Ag^+ + 2NH_3$ is 7.2; log K_1 for $Ag^+ +$ en is around 6). For Hg^{2+} complexing with the edta homologues shown in the lower half of Table 6.6, log K_1 is a *minimum* for the six-membered ring, rising gradually as the chelate ring increases in size up to eleven atoms.

6.4 SELECTIVITY; MACROCYCLIC AND ENCAPSULATING LIGANDS

In the previous section we saw how the chelate effect and, to an even greater extent, the macrocyclic effect, could give rise to particularly stable complexes. In this section we continue this theme, showing how selectivity can be built into macrocyclic ligands and how even labile metal ions can be encapsulated or caged to give very stable and inert complexes. These topics lead from standard inorganic chemistry into bioinorganic and *in vivo* systems.

Electropositive elements such as the lanthanides and the alkaline earths form relatively few complexes, and these are of relatively low stability. The highly electropositive alkali metals are especially reluctant to form complexes. Ligands need to contain several 'hard' donor atoms, such as nitrogen or, better, oxygen, and to be of such a structure that the chelate or macrocyclic effects can be brought into full play. Even polydentate ligands of the edta type and the four-nitrogen macrocycles encountered in the previous section, which form very stable complexes with transition metal cations, prove to be poor ligands for the alkali metal cations. However, crown ethers, which are oxygen analogues of the tetra-aza-macrocycles (right-hand side of Fig. 6.5), do form complexes with the alkali metal cations. Fig. 6.6 shows examples of this type of ligand, with an indication of their universally used trivial names. Crown ether complexes of the alkali metal cations are often stable in non-aqueous media, rarely in water. The slightly more complicated cryptand family of ligands, Fig. 6.7, form somewhat more stable complexes, some of which are stable in aqueous media.

The ligands shown in Figs. 6.6 and 6.7 are of considerable interest simply as effective complexing agents for alkali metal cations. Indeed they are of synthetic value in organic chemistry, as they convert hydrophilic salts alk$^+$X$^-$ into salts of the type [alk.crown]$^+$X$^-$ which are soluble in many organic solvents. This enables one to carry out substitutions by such ions as fluoride or cyanide rather easily. Such reactions are often fast, since the inorganic anion is generally weakly solvated or unsolvated ('naked fluoride') and thus at a high chemical potential and an aggressive nucleophile. Beyond this powerful complexing ability, this group of ligands is of great interest in their powers of selectivity. They wrap themselves around a metal ion so that as many donor atoms as possible can bond to the cation. An example of this is shown in Fig. 6.8(a). The ligand donor atoms effectively form a cavity to hold the metal ion, with the size of this cavity determined by the precise structure and

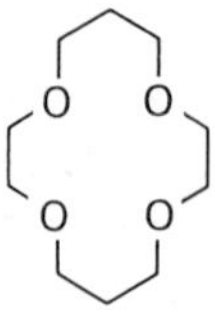

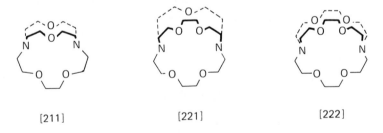

14-CROWN-4 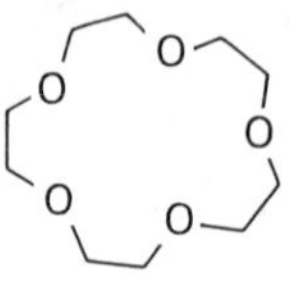   15-CROWN-5 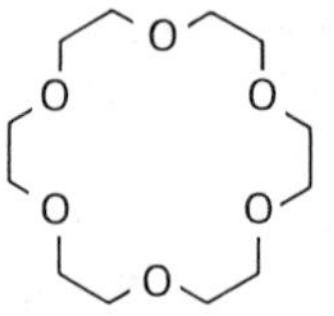   16-CROWN-6

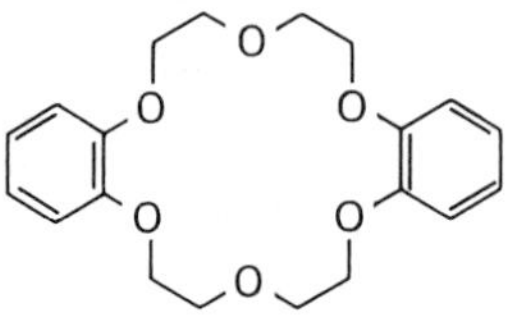

DIBENZO-18-CROWN-6 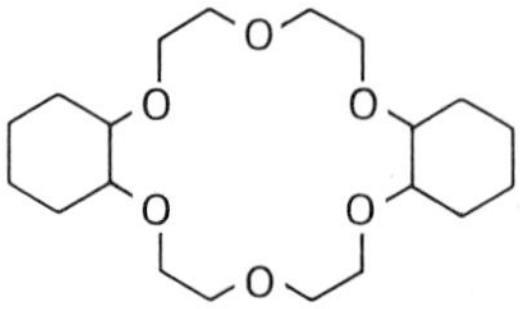   DICYCLOHEXYL-18-CROWN-6

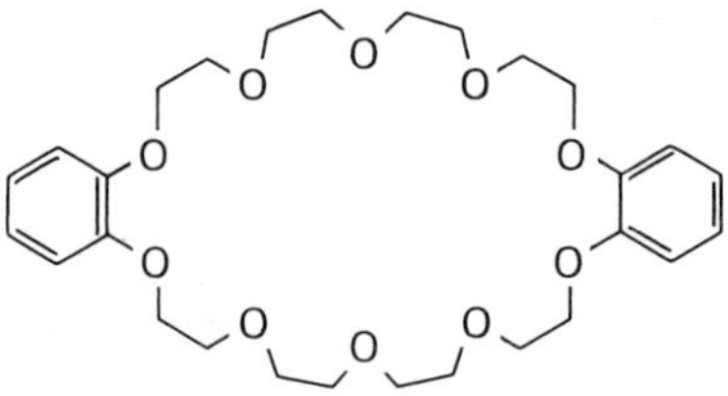

DIBENZO-30-CROWN-10

Fig. 6.6 — Crown ethers, with their conventional names.

[211]   [221]   [222]

Fig. 6.7 — Cryptands, with their conventional designations (the numbers within the square brackets indicate the number of oxygen atoms in each $\{CH_2CH_2O\}_n$ link).

dimensions of the ligand. The radii of these cavities are of comparable magnitude to the ionic radii of metal cations (Table 6.7) which is where the selectivity comes in. If the cation is rather too small for the cavity, it is unable to interact properly with all the

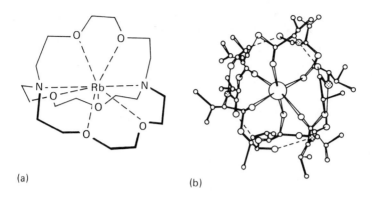

(a)   (b)

Fig. 6.8 — (a) Encapsulation of Rb$^+$ by the cryptand [222]; (b) encapsulation of K$^+$ by valinomycin.

Table 6.7 — Relation of ionic radii to cavity size for crown ether and cryptand ligands

	Cavity^a radius (Å)		Cf. ionic^b radius (Å)
Crown ethers			
12-crown-4	ca. 0.6		
15-crown-5	0.9	Li$^+$	ca. 0.8^c
18-crown-6	1.4	Na$^+$	1.16
21-crown-7	1.7	K$^+$	1.52
Cryptands		Rb$^+$	1.63
[211]	0.8	Cs$^+$	1.84
[221]	1.1^d		
[222]	1.4		
[322]	1.8		
[332]	2.1		
[333]	2.4		

^{a}From molecular models and X-ray diffraction.
^{b}Shannon and Prewitt values, for 6-coordination.
^{c}The ionic radius of Li$^+$ varies considerably according to the method used for assigning ionic radii.
^{d}Na$^+$ fits the [221] cavity well, K$^+$ is too big; the mean Na$^+$–O distance in the thiocyanate salt of Na[221]$^+$ as determined by X-ray diffraction is 2.48 Å, which can be compared with the sum of the ionic radius of Na$^+$ and the cavity radius, ≈ 2.3 Å.

potential donor atoms. If it is rather too large, then work will have to be done to enlarge the cavity. Either way the resulting complexes will be markedly less stable than one in which the cation fits snugly in the cavity, interacting to the maximum extent with each donor atom. In general as one goes down a group of cations, such as the $Li^+ \rightarrow Na^+ \rightarrow K^+ \rightarrow Rb^+ \rightarrow Cs^+$ sequence, one might expect an increase followed by a decrease in stability as the ionic radius rises to, and then beyond, the optimum value for the cavity in the ligand in question. Such behaviour is illustrated in Fig. 6.9 for a series of cryptands (Fig. 6.7) and their complexes.

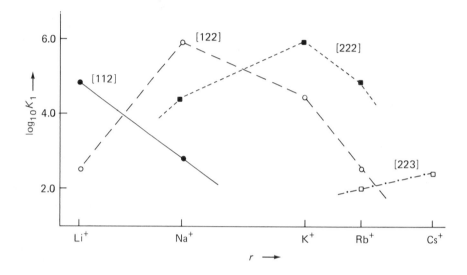

Fig. 6.9 — Stability constant trends for cryptates of alkali metal cations.

The biochemical relevance of this lies in the selective transport of sodium and potassium ions across cell membranes, maintaining the different concentrations of these cations inside and outside cells. The complexing agents found *in vivo* are crown ethers, of which some examples are shown in Fig. 6.10. The complexing function (see e.g. Fig. 6.8(b)) is very similar to the ligands shown in Fig. 6.6; other parts of the molecule are more complicated to fit these species for their role in the living organism. Enniatin B (Fig. 6.10(a)) has an 18-membered ring, and complexes to K^+ through six carbonyl oxygens; the parallel to 18-crown-6 is very close. The cation selectivity of three of these bio-ligands is compared with the similar selectivity of 18-crown-6 in Table 6.8, albeit in non-aqueous solvents rather than in a biological medium.† In each case there is marked discrimination between Na^+ and K^+. Interestingly, nearly all the biochemical ligands selectively bind K^+ in preference to Na^+; only the monensin anion shows a marked preference for Na^+. Besides its biological importance, valinomycin (Fig. 6.10(b)) is also of analytical importance in an ion selective electrode for K^+.

---

† See Chapter 13 for mention of recent work in systems very much more closely related to *in vivo* cation transport.

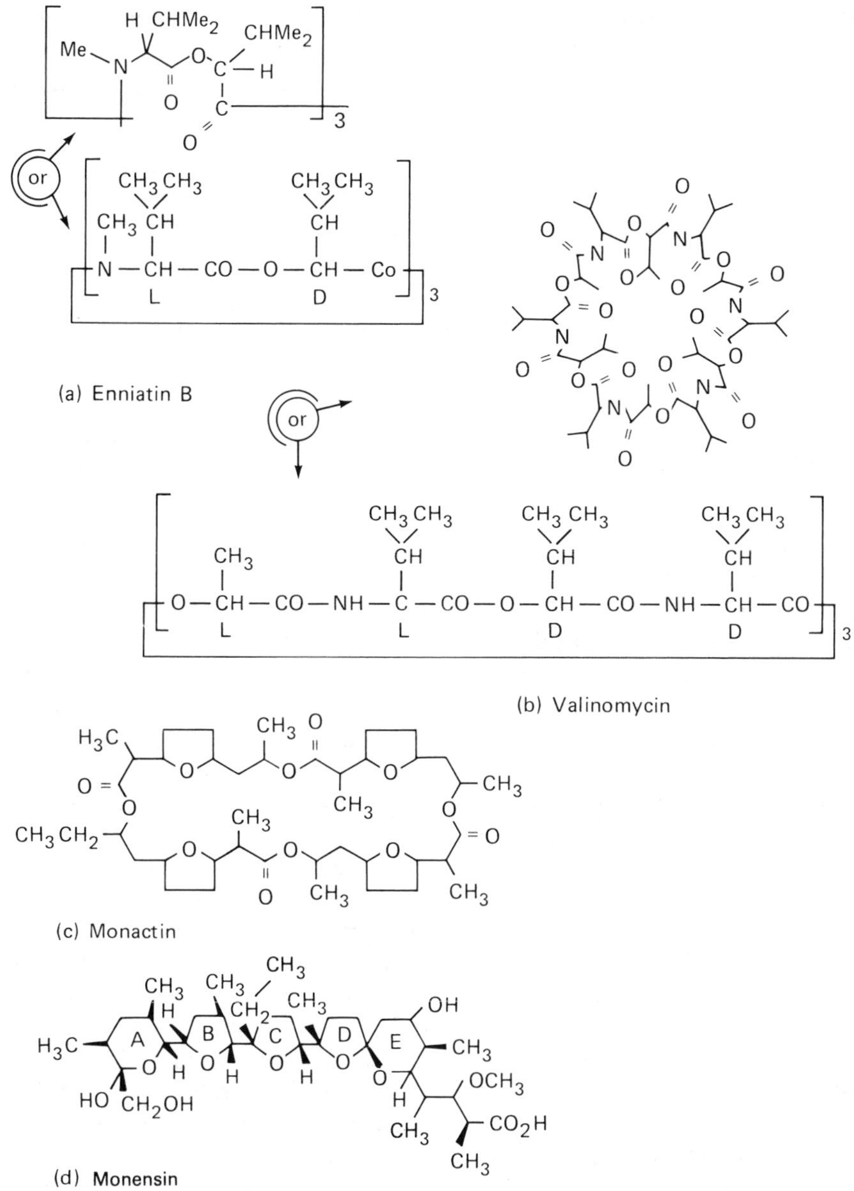

(a) Enniatin B

(b) Valinomycin

(c) Monactin

(d) Monensin

Fig. 6.10 — Some biological crown ether ligands (monensin, (d), cyclises by hydrogen-bonding between its terminal OH and CO_2H groups).

Stability constants represent the balance between forward and reverse reactions in equilibria

$$M^{n+} + L \underset{k_b}{\overset{k_f}{\rightleftharpoons}} ML^{n+}$$

In Chapters 10 and 11 we shall discuss the kinetics and mechanisms of formation and of dissociation of metal complexes, including those of crown ethers and cryptands.

Table 6.8 — Stability constants ($\log_{10}K_1$) for formation of complexes between alkali metal cations and crown ethers

Ligand:	18-crown-6		valinomycin	monactin	monensin[a]
Solvent:	acetone	propylene carbonate	methanol	methanol	methanol
Li^+	1.5	2.7	v. low	v. low	3.6
Na^+	4.6	5.2	2.6	2.3	6.5
K^+	6.0	6.1	4.8	3.6	5.2
Rb^+	5.2	5.2	5.1	3.5	4.5
Cs^+	4.6	4.5	4.3	2.9	3.7

[a]Monensin behaves as a crown ether as its two ends are linked by hydrogen bonding; see Fig. 6.10(d).

For this class of complexes, it is found that for a given metal ion k_f varies rather little with the nature of the ligand. Variations in stability constants, i.e. selectivities, are controlled by k_b, by dissociation rather than by formation. Cryptates and crown ether complexes are formed relatively slowly, but dissociate even more slowly. In dissociation it is necessary not only to break several metal–ligand bonds but also to invert at several atoms in order to change the confirmation of the ligand sufficiently to allow the metal ion to escape. We shall mention these relations between stability and rate constants for formation and dissociation again in Chapter 8, in connection with the existence or otherwise of correlations between kinetic and thermodynamic parameters.

The preceding paragraphs have concentrated on the alkali metal cations, but crown ethers, cryptands, and related macropolycyclic ligands are also effective at complexing a variety of other metal ions. In particular they form stable complexes with Group II cations (Mg^{2+}, Ca^{2+}, Sr^{2+}, Ba^{2+}), again showing marked selectivity controlled by the relation between cavity size and cation ionic radius. Table 6.9 shows the enormous favouring of Ba^{2+} and Sr^{2+} over the smaller Ca^{2+} and Mg^{2+} by the cryptand [222],†. which has a large cavity, and the maximum in the stability constant sequence at Sr^{2+} for the azacrown ligand shown in the table, which has a slightly smaller cavity. The Tl^+ ion has the same radius as Rb^+, and forms crown ether and cryptate complexes of comparable stability. Ag^+ and Pb^{2+} have ionic radii intermediate between those of Na^+ and K^+, and can be transported by monensin in a biochemical environment. Monensin is ineffective in transporting 2 + ions of the transition metals; their ionic radii are all < 1 Å and therefore too small for the cavity available.

This approach of encapsulating or caging a metal ion to give a complex of unusual inertness and high stability has also been applied to transition metal cations. In contrast to the crown ether and cryptate complexes, which are formed by adding the preformed (and often commercially available) ligand to a solution containing a metal

† The cryptand [222] complexes Ba^{2+} so strongly that a litre of a concentrated aqueous solution of this ligand is said to be able to dissolve up to about 50 g of barium sulphate.

Table 6.9 — Stability constants ($\log_{10}K_1$, molar scale, 298.2 K) for complexes of Group II cations with cryptand and azacrown ether ligands

	in water	in methanol
Mg^{2+}	2.0	3.8
Ca^{2+}	4.1	4.3
Sr^{2+}	13.0	<u>6.1</u>
Ba^{2+}	>15	5.9

ion, most transition metal analogues are prepared by template syntheses. A relatively simple complex is treated with appropriate reagents to generate the cage around the metal ion. One of the earliest and simplest examples was the reaction of the tris-ethane-1,2-diamine complex $[Co(en)_3]^{3+}$ with formaldehyde and ammonia in basic solution to give the so-called sepulchrate complex $[Co(sep)]^{3+}$. This is shown in Fig. 6.11(a) and (b), alongside Werner's hexol, Fig. 6.11(c), whose central cobalt is arguably the earliest caged metal ion. This sepulchrate complex is even more inert than 'normal' cobalt(III) complexes, but what is more striking is the great inertness of its cobalt(II) analogue, $[Co(sep)]^{2+}$. This undergoes negligible exchange with radioactive Co^{2+} aq in a day at 298 K, in contrast to the usual timescale of the order of milliseconds or less for substitution at Co^{2+}. Condensation of the tris-dimethyl-glyoxime complex of iron(II) with boric acid in butan-1-ol gives a similar caged complex (Fig. 6.12(a)). Boron trifluoride, conveniently used in the form of its etherate, gives an analogous complex with a pair of $>$BF caps (Fig. 6.12(b)) instead of $>$BO"Bu, while condensation of tris-dihydrazone complexes with formaldehyde gives triazacyclohexane capping (Fig. 6.12(c)). Such very inert complexes, and related species such as the $[CoW_{12}O_{40}]^{5-}$ anion (see next paragraph), are valuable in studies of kinetics and mechanisms of electron transfer reactions (Chapter 12). When both the oxidised and reduced forms of a complex are so inert there is no possibility of complications arising from ligand dissociation.

Ths $[CoW_{12}O_{40}]^{5-}$ anion just mentioned is one of a large family of heteropoly-tungstates and heteropolymolybdates. Many of these species contain a metal ion at the centre of a framework of linked WO_4 or MoO_4 tetrahedra. The $[MW_{12}O_{40}]^{2-}$ series can have M as, for example, Fe^{III}, Cr^{III}, Cu^{II}, Co^{II}, as well as Co^{III}. The M^n ion is in tetrahedral coordination, surrounded by a symmetrical framework (the tungsten atoms are at the corners of a cuboctahedron) of linked WO_4 tetrahedra. This is the so-called 'Keggin structure', named after the man who first established it. Such complicated structures are very difficult to represent clearly in two dimensions; Fig. 6.13 shows four aspects (a–d). The four groups of fused WO_4 tetrahedra can be seen

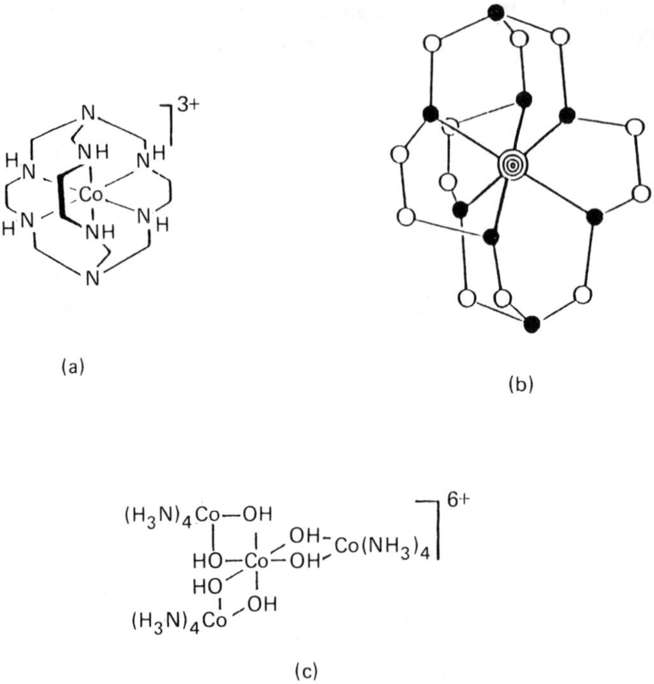

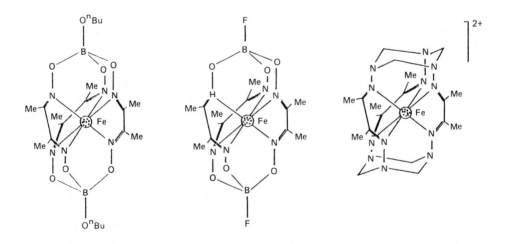

Fig. 6.11 — (a), (b) Two representations of the cobalt(III) sepulchrate complex; (c) Werner's
hexol cation.

Fig. 6.12 — Iron(II) complexes of encapsulating (cage) ligands.

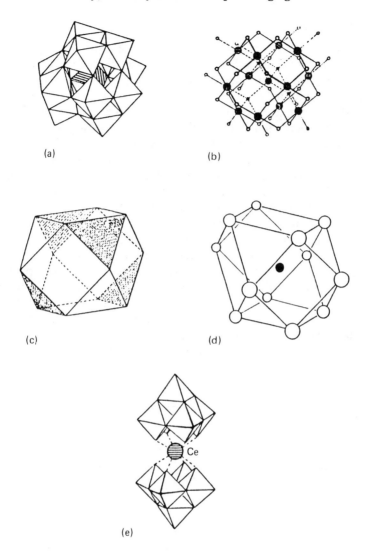

Fig. 6.13 — (a), (b), (c), (d) Four representations of the Keggin structrue for heteropoly tungstates $[MW_{12}O_{40}]^{n-}$; (e) the $[CeW_{10}O_{36}H_2]^{6-}$ anion.

in Fig. 6.13(a), while Fig. 6.13(b) shows the M atom at the centre and the surrounding skeleton of W and O atoms. Fig. 6.13(c) shows a cuboctahedron, with the shaded faces indicating the positions of the four groups of WO_4 tetrahedra displayed in Fig. 6.13(a). Fig. 6.13(d) shows the same cuboctahedron as Fig. 6.13(c), but this time with the twelve tungsten atoms and the central metal atom specifically indicated. Finally 6.13(e) gives an alternative picture of the shielding of a central metal ion from the outside world for the closely related case of the $[CeW_{10}O_{36}H_2]^{6-}$ anion, where the less enclosed nature of the cerium permits easier drawing.

6.5 RELEVANCE

Stability constants are important in many fields of chemistry — analytical, biological, domestic, and industrial as well as inorganic. Thus the use of ethane-1,2-diamine-tetraacetate (edta) in complexometric titrations depends both on the high stabilities of its complexes with many metal ions and on appropriate balances between the stabilities of the complexes of the metal ions with edta on the one hand and the various complexometric indicators on the other. Biological examples include stability constants for complexes of haemoglobin and of myoglobin with dioxygen, carbon monoxide, and cyanide. Here the absolute and relative stabilities ensure that the dioxygen complex of haemoglobin is formed in the lungs, but that this dioxygen dissociates at the site where it is needed, permit the resuscitation of a person in the early stages of carbon monoxide poisoning, but totally preclude the displacement of cyanide by dioxygen. Domestic relevance can be illustrated by the use of silver-cleaning solutions, which contain a ligand with a higher affinity for Ag^+ than the tarnish, and by the use of various iron(III) complexing agents in rust removers. The latter is also of relevance industrially, as are, for instance, the polyphosphates and polyaminocarboxylates of the edta type used to sequester traces of unwanted metal ions. Pharmaceutical applications include the use of a range of ligands to remove toxic metals. Here again it is important to consider relations between stability constants for various metal ions, for one must avoid depleting the body content of Mg^{2+}, Ca^{2+}, and Zn^{2+} when one attempts to remove traces of Cd^{2+}, Hg^{2+}, Pb^{2+}, or Pu^{3+}.

7

Redox potentials

7.1 INTRODUCTION AND THERMODYNAMICS

The next fundamental reaction of ions in solution to be considered is that of electron transfer to or from the ion, reduction or oxidation respectively:

$$M^{3+} + e^- \underset{\text{oxidation}}{\overset{\text{reduction}}{\rightleftharpoons}} M^{2+} \qquad (7.1)$$

$$M^+ + e^- \underset{\text{oxidation}}{\overset{\text{reduction}}{\rightleftharpoons}} M^0 \qquad (7.2)$$

Such reactions are feasible at electrodes, or in the presence of hydrated electrons. Normal homogeneous solution reactions can be made up from pairs of such 'half-reactions':

$$Cu^{2+} + 2e^- = Cu^0$$
$$Zn^{2+} + 2e^- = Zn^0$$

Combine: $\quad Cu^{2+} + Zn^0 = Cu^0 + Zn^{2+}$

This is equivalent to coupling two such immersed electrodes or half-cells (Fig. 7.1). The reaction involved here can be carried out in a single vessel by immersing a piece of metallic zinc in copper(II) sulphate solution, to give the 'copper tree'.

The electron transfer processes outlined above can be treated thermodynamically, with equation (7.3) forming

$$\Delta G^\ominus = -n\mathcal{F}E^\ominus \qquad (7.3)$$

the link between the Gibbs free energy for electron transfer to the electrochemically

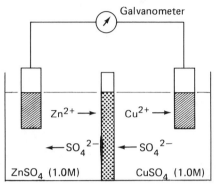

Fig. 7.1 — A Daniell cell.

measurable potential for an n-electron change. Half-cell redox potentials for hypothetical reactions of the type shown in equations (7.1) and (7.2) are valuable in inorganic chemistry in two ways. Firstly, they simply provide a quantitative measure of the ease of oxidising or reducing a given species in solution. Secondly, by taking $E^\ominus$ values in appropriate pairs it is possible to derive $\Delta G^\ominus$ values for redox reactions. The sign of $\Delta G^\ominus$ gives the fundamental information on the direction of spontaneous reaction; if $\Delta G^\ominus$ is close to zero, then the equilibrium composition reached on reacting oxidant and reductant can be calculated in the usual way (cf. equation (7.4)).

$$\Delta G^\ominus = - RT \ln K \qquad (7.4)$$

Half-cell $E^\ominus$ values are a highly efficient way to store redox information; a table of only 20 values contains $\Delta G^\ominus$ for $20 \times 19 = 380$ reactions. Against all these advantages one should remember to set certain restrictions on the usefulness of $E^\ominus$ values. The values quoted in most tabulations refer to aqueous solution, unit activity of hydrogen ions, and 298.2 K. Data under other conditions are available but very sparse. It is impossible to compare $E^\ominus$ values in different solvent media without involving an extrathermodynamic assumption — and indeed it should be borne in mind that the universally used scale of standard $E^\ominus$ values is based on an arbitrary zero, the potential of the standard hydrogen electrode.† Finally, redox potentials and $\Delta G^\ominus$ values derived from them are thermodynamic and not kinetic properties. They give information about final states, not about the rates of attaining equilibrium. Solutions

---

† In fact the absolute value of the potential for the standard hydrogen electrode is believed to be approximately $+4.5$ V, corresponding to a Gibbs free energy of hydration of the proton of -1086 kJ mol^{-1} (cf. the enthalpy of hydration of the proton of -1091 kJ mol^{-1} given in section 4.3).

of permanganate or of peroxodisulphate are thermodynamically unstable, but take a very long time to decompose!

7.2 AQUA-METAL IONS

A selection of redox potentials for inorganic half-cell reactons is given in Table 7.1;

Table 7.1 — Redox potentials for inorganic half-cells

Oxidant	Reductant	$E^\ominus$(V)
Powerful oxidants		
$F_2 + 2e^-$	$= 2F^-$	2.87
$S_2O_8^{2-} + 2e^-$	$= 2SO_4^{2-}$	2.12
$MnO_4^- + 8H^+ + 5e^-$	$= Mn^{2+} + 4H_2O$	1.51
$Cl_2 + 2e^-$	$= 2Cl^-$	1.36
$O_2 + 4H^+ + 2e^-$	$= 2H_2O$	1.23
$Br_2 + 2e^-$	$= 2Br^-$	1.09
$Ag^+ + e^-$	$= Ag$	0.80
$Fe^{3+} + e^-$	$= Fe^{2+}$	0.77
$Cu^{2+} + 2e^-$	$= Cu$	0.34
$2H^+ + 2e^-$	$= H_2$	0
$Cr^{3+} + e^-$	$= Cr^{2+}$	-0.41
$Zn^{2+} + 2e^-$	$= Zn$	-0.76
$Al^{3+} + 3e^-$	$= Al$	-1.66
$Mg^{2+} + 2e^-$	$= Mg$	-2.37
$Na^+ + e^-$	$= Na$	-2.71
Powerful reductants		

some uses of such redox potentials are illustrated in Figs 7.2 to 7.4. The diagram summarising the aqueous redox chemistry of manganese (Fig. 7.4) contains two examples of the relatively rare phenomenon of disproportionation (Fig. 7.2(c)), otherwise important only for uranium, neptunium, plutonium, and copper. It is interesting to note that although copper(I) is unstable with respect to disproportionation into copper(II) plus copper metal in aqueous media, copper(I) is stable to disproportionation in acetonitrile or in ethanol. Copper(I) is marginally stable in dimethyl sulphoxide, where the equilibrium constant for the reacton

$$2Cu^+ \rightleftharpoons Cu^{2+} + Cu^0$$

has the value 2 (in water it is 1.7×10^6).

Having dealt in some detail with the redox potential diagrams for vanadium and for manganese, it is now time to examine some trends of redox potentials in relation to the Periodic Table. Table 7.2 shows how $E^\ominus$ values vary on descending Groups I

(a) *Direction of reaction*

$$Cu^{2+} + 2e^- = Cu^0 \quad E^\ominus = +0.337 \text{ V} \qquad \Delta G^\ominus = -n\mathcal{F}E^\ominus = -65.0 \text{ kJ mol}^{-1}$$
$$Zn^{2+} + 2e^- = Zn^0 \quad E^\ominus = -0.763 \text{ V} \qquad \Delta G^\ominus = -n\mathcal{F}E^\ominus = +147.2 \text{ kJ mol}^{-1}$$

subtract $Cu^{2+} - Zn^{2+} = Cu^0 - Zn^0$ $\qquad\qquad\qquad\qquad \Delta G^\ominus = -82.2 \text{ kJ mol}^{-1}$
rearrange $Cu^{2+} + Zn^0 = Cu^0 + Zn^{2+}$

Conclusion: Reaction spontaneous left → right.

(b) *Equilibrium*

$$Cr^{3+} + e^- = Cr^{2+} \quad E^\ominus = -0.41 \text{ V} \qquad \Delta G^\ominus = +39.6 \text{ kJ mol}^{-1}$$
$$Eu^{3+} + e^- = Eu^{2+} \quad E^\ominus = -0.35 \text{ V} \qquad \Delta G^\ominus = +33.8 \text{ kJ mol}^{-1}$$

hence $Cr^{3+} + Eu^{2+} = Cr^{2+} + Eu^{3+}$ $\qquad\qquad\qquad \Delta G^\ominus = +5.8 \text{ kJ mol}^{-1}$

Conclusion: Small driving force in the reverse (right → left) direction;
equilibrium constant = 0.1 from $\Delta G^\ominus = -RT \ln K$.

(c) *Disproportionation*
$$Mn^{IV} + e^- = Mn^{3+} \quad E^\ominus = +0.95 \text{ V} \qquad \Delta G^\ominus = -91.7 \text{ kJ mol}^{-1}$$
$$Mn^{3+} + e^- = Mn^{2+} \quad E^\ominus = +1.51 \text{ V} \qquad \Delta G^\ominus = -145.7 \text{ kJ mol}^{-1}$$

hence $2Mn^{3+} = Mn^{2+} + Mn^{IV}$ $\qquad\qquad\qquad\qquad \Delta G^\ominus = -54.0 \text{ kJ mol}^{-1}$

Conclusion: Disproportionation of manganese(III) to manganese(II) plus manganese(IV) is thermodynamically favourable.

Fig. 7.2 — Redox potentials and Gibbs free energy changes: (a) direction of reaction; (b) equilibrium; (c) disproportionation.

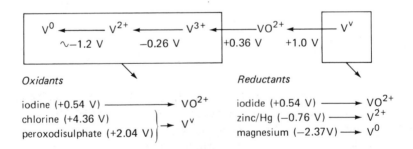

Fig. 7.3 — Redox potential diagram for vanadium.

and II of the Periodic Table, for the M^+/M^0 and M^{2+}/M^0 complexes respectively. The Group II trend is simple, but the Group I trend shows a slight extremum, suggesting that more than one factor determines the values. The factors which contribute are set out in Fig. 7.5. Table 7.3 lists the values for the enthalpies for the various components, and shows how a non-simple trend of $E^\ominus$ values can arise ($T\Delta S^\ominus$ contributions may also affect trends). Table 7.3 shows, for example, how the balance

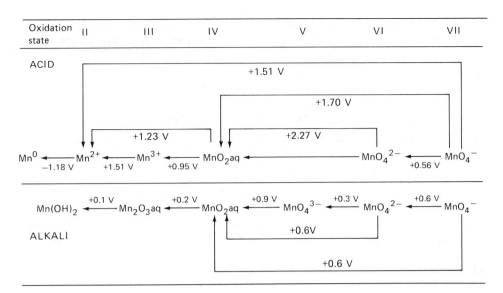

Fig. 7.4 — Redox potential diagram for manganese.

Table 7.2 — Redox potentials for the alkali and alkaline earth metals

	$E^{\ominus}(V)$		$E^{\ominus}(V)$
Li^+/Li	-3.05	Be^{2+}/Be	-1.85
Na^+/Na	-2.71	Mg^{2+}/Mg	-2.37
K^+/K	-2.93	Ca^{2+}/Ca	-2.87
Rb^+/Rb	-2.93	Sr^{2+}/Sr	-2.89
Cs^+/Cs	-2.92	Ba^{2+}/Ba	-2.90
		Ra^{2+}/Ra	-2.92

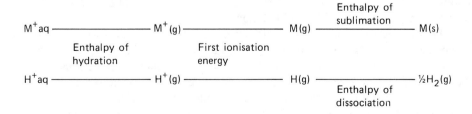

Fig. 7.5 — Factors contributing to redox potentials for half-cells.

Table 7.3 — Analysis of redox potentials into their component contributions — enthalpies of sublimation (or dissociation), ionisation enthalpies, and ion hydration enthalpies (for M^{n+}/M; all in kJ mol^{-1})

	H	
$\frac{1}{2}H_2(g) \rightarrow H(g)$		$+218$
$H(g) \rightarrow H^+(g) + e^-$		$+1318$
$H^+(g) \rightarrow H^+(aq)$		-1091
		$+445$

	Li	Na	K	Rb	Cs	Ag
$M(s) \rightarrow M(g)$	$+161$	$+108$	$+90$	$+82$	$+78$	$+286$
$M(g) \rightarrow$ $M^+(g) + e^-$	$+526$	$+502$	$+425$	$+409$	$+382$	$+737$
$M^+(g) \rightarrow$ $M^+(aq)$	-515	-405	-326	-296	-263	-475
	$+172$	$+205$	$+189$	$+195$	$+197$	$+540$

	Ca	Cu	Zn
$M(s) \rightarrow M(g)$	$+193$	$+339$	$+131$
$M(g) \rightarrow$ $M^{2+}(g) + 2e^-$	$+1735$	$+2715$	$+2651$
$M^{2+}(g) \rightarrow$ $M^{2+}(aq)$	-1592	-2100	-2044
	$+336$	$+954$	$+738$

between ionisation energies and hydration energies, both decreasing down Group 1A, can give an irregular $E^{\ominus}$ sequence, and how the very much greater energy of vaporisation of silver metal than of the alkali metals is an important factor in determining $E^{\ominus}$ (Ag$^+$/Ag^0). By way of contrast the redox potentials for the halogen couples $\frac{1}{2}X_2/X^-$ decrease steadily down the Periodic Table, with $E^{\ominus}$ for X = F +2.87 V, Cl +1.36 V, Br +1.07 V, and I +0.54 V.

The variation of redox potentials for the M^{3+}/M^{2+} couples for the first-row d-block elements is shown in Fig. 7.6. All the factors detailed in Fig. 7.5 and Table 7.3 are relevant, as well as the crystal field stabilisation energies of the various M^{2+} and M^{3+} ions. The drop in $E^{\ominus}$ on going from Mn to Fe can be traced to the variation in $M^{2+} \rightarrow M^{3+}$ ionisation energies (gas phase), while the overall trend across the whole series reflects a steady decrease in the stability of the higher oxidation state with respect to the lower from Sc^{3+} through to Ni^{3+}. Fig. 7.7 shows the variation of redox potentials for the M^{3+}/M^{2+} couples for the first row of the f-block elements, the

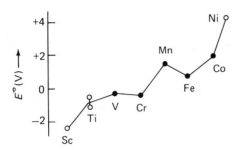

Fig. 7.6 — The variation of the M^{3+}/M^{2+} redox potentials along the first row of the transition metals; ● represents an accurate value, ○ represents an approximate estimate.

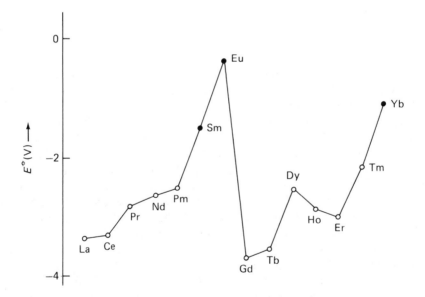

Fig. 7.7 — The variation of the M^{3+}/M^{2+} redox potentials along the first series of f-block elements (the lanthanides); ● represents an accurate value, ○ represents an approximate estimate.

lanthanides. Here the majority of values are estimates, owing to the instability of most lanthanide(II) cations in aqueous media. However, this representation shows very clearly the marked stabilisation of Eu^{2+} and Yb^{2+}, whose f^7 and f^{14} configurations (half-filled and filled $4f$ shells) confer exceptional stability (cf. the complementary case of the f^0 and f^7 ions Ce^{4+} and Tb^{4+}).

7.3 COMPLEXES

All the redox potentials discussed so far have related to aqua-ions $M(OH_2)_x^{n+}$. Replacing one or more of the water molecules by other ligands has a marked effect on redox potentials. This can be illustrated by the Ce^{4+}/Ce^{3+} couple. In order to prevent hydrolysis and polymerisation (Chapter 5), cerium(IV) solutions are made up in strongly acidic media. The $E^{\ominus}$ value for Ce^{4+}/Ce^{3+} is found to vary considerably with the nature of acid used (Table 7.4); cerium(IV) is a much weaker oxidant in

Table 7.4 — Effect of the nature of acid used to maintain $a(H^+) = 1$ on the Ce^{4+}/Ce^{3+} redox potential

Perchloric acid	+ 1.70 V
Nitric acid	+ 1.61 V
Sulphuric acid	+ 1.44 V
Hydrochloric acid	+ 1.28 V

hydrochloric acid than in perchloric acid. These effects of complex formation between cerium(IV) and the anions of the acids used (except perchloric) are important in some analytical applications.

Perhaps the best documented couple in respect of effects of ligand variation on redox potential is Fe^{3+}/Fe^{2+}. A selection of the available data is shown in Fig. 7.8. Column I gives an idea of the range of values; it is interesting to compare this range of ca. 1.5 V with the range of $E^{\ominus}$ values for, e.g., the M^{2+}/M^0 couples for Group IIA (Table 7.2) or the M^{3+}/M^{2+} couples of d-block elements (Fig. 7.6). Such comparisons show the large effect of ligands on redox behaviour. Several factors contribute to this. Inspection of column I of Fig. 7.8 reveals that complexes containing anionic ligands occur towards the bottom; anions tend to stabilise Fe^{3+} more than Fe^{2+} on simple electrostatic grounds. Conversely ligands such as 2,2′-bipyridyl and 1,10-phenanthroline (Fig. 7.9) stabilise Fe^{2+} more than Fe^{3+} (Fig. 7.10); this π-electron transfer is easier from a 2+ (d^6) than a 3+ (d^5) ion. Cyanide is both negative and a good π-acceptor, so it is reassuring to find $E^{\ominus}$ ($Fe(CN)_6^{3-}/Fe(CN)_6^{4-}$) in the middle of column I. The other columns of Fig. 7.8 give an idea of the range and variation of redox potentials produced by changing ligands in various other series, illustrating the possibilities of fine-tuning (columns II and III) and some bioinorganic systems (columns III and IV).

Another couple whose redox potential dependence on ligand nature is of great importance is Co^{3+}/Co^{2+} (Table 7.5). Co^{3+} aq is one of the few high-spin cobalt(III) complexes. Its instability is reflected in the high value for the uncomplexed Co^{3+}/Co^{2+} couple in aqueous solution. Oxygen-donor anionic ligands such as oxalate stabilise the cobalt(III) sufficiently for $[Co(ox)_2(OH_2)_2]^-$ and $[Co(ox)_3]^{3-}$ to be stable in aqueous solution, in contrast to Co^{3+} aq which quickly oxidises water. Chelating anionic ligands with oxygen and nitrogen donor ligands, e.g. edta, cause further stabilisation of the 3+ oxidation state, and significantly greater stabilisation

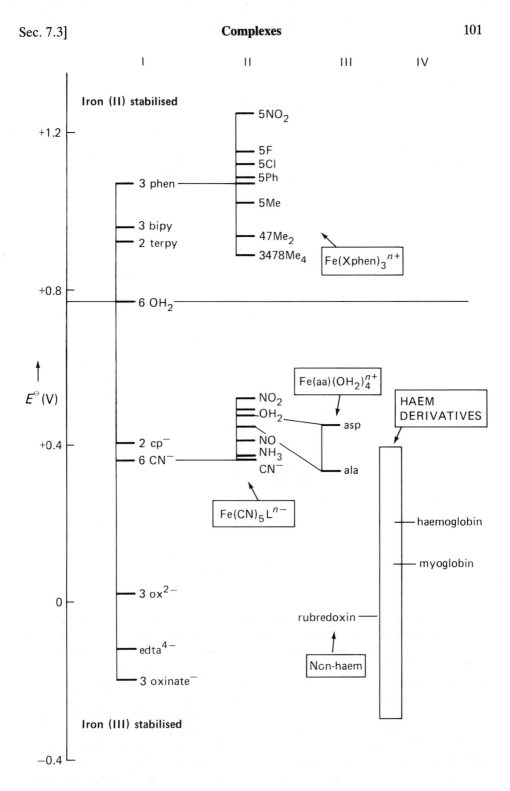

Fig. 7.8 — The effect of complex formation on the iron/iron couple (in all cases in aqueous solution).

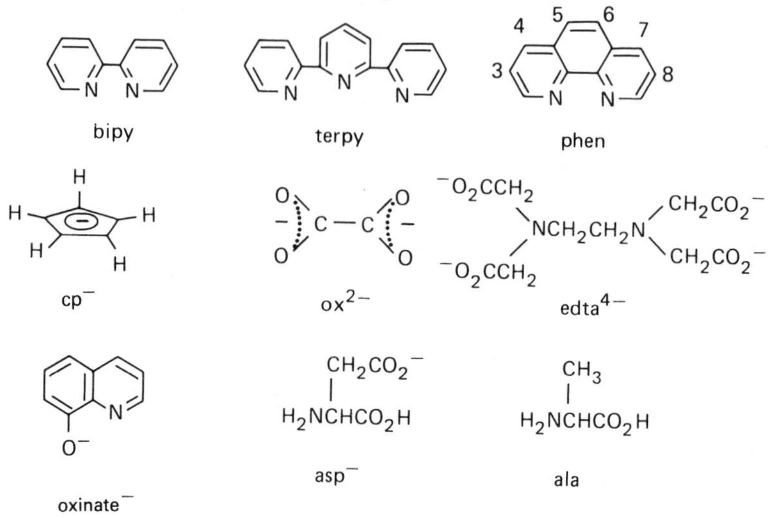

bipy

terpy

phen

cp^-

ox^{2-}

$edta^{4-}$

oxinate$^-$

asp^-

ala

Fig. 7.9 — Ligand formulae for Fig. 7.8.

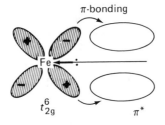

Fig. 7.10 — Bonding in low-spin d^6 complexes.

of this higher oxidation state is achieved by the coordination of a complete set of nitrogen donors, e.g. six ammonia or three ethane-1,2-diamine ligands. Finally, the cyano-complexes couple has a large negative redox potential, reflecting considerable stabilisation of the 3+ oxidation state over the 2+ state. The negative charge of cyanide and its strong π-acceptor properties combine with the large crystal field stabilisation for a t_{2g}^6 ion to result in this strong preferential stabilisation of the cobalt(III) state. The steadily decreasing $E^\ominus$ values going down Table 7.5 reflect steadily increasing stabilisation of the cobalt(III) state; the cobalt(II) remains high-spin throughout. In the couples at the bottom of the table, the cobalt(II) is unstable with respect to cobalt(III) in the presence of air. If one takes a ligand that greatly stabilises cobalt(II) — for instance the hexathiamacrocycle shown in Fig. 7.11, which makes even the cobalt(II) low-spin — then this stabilisation is reflected in a high $E^\ominus$

Table 7.5 — Effects of ligands on the cobalt(III)/(II) redox couple

	$E^{\ominus}(V)$
$Co^{3+}aq/Co^{2+}aq$	$+1.88$
$[Co(ox)_2(OH_2)_2]^{-/2-}$	$+0.78$
$[Co(ox)_3]^{3-/4-}$	$+0.57$
$[Co(edta)]^{-/2-}$	$+0.37$
$[Co(bipy)_3]^{3+/2+}$	$+0.31$
$[Co(en)_3]^{3+/2+}$	$+0.18$
$[Co(NH_3)_6]^{3+/2+}$	$+0.11$
$[Co(CN)_6]^{3-}/[Co(CN)_5]^{3-},CN^-$	-0.8
cf. $\frac{1}{2}O_2, 2H^+/H_2O$	$+1.23$

Fig. 7.11 — Hexathiamacrocyclic ligand.

value for the Co^{III}/Co^{II} redox potential (about $+1\,V$). It is interesting to compare this hexathiamacrocycle couple, where the strong ligand field stabilises both the 2+ and 3+ states, with the hexaaqua-couple, where both oxidation states are high-spin due to the weak ligand field of the water ligands.

Some redox potential data for copper(II)/(I) couples are included in Fig. 7.12, where they are compared with a selection of the iron(III)/(II) and cobalt(III)/(II) couples already mentioned. This figure shows how much the effects of ligands on redox potentials depend on the nature and specific chemical properties of the metal ions and of the ligands involved. In the case of the copper(II)/(I) couple there is the extra complication of the distorted octahedral/tetrahedral change in geometry involved in redox reactions. The iron(III)/(II) and cobalt(III)/(II) couples do not involve changes in geometry or coordination number, except in the case of the Co(III)/(II)/CN$^-$ couple. Several of the copper(II)/(I) couples for which redox potential data are available involve bioinorganic materials. Often such materials contain the copper complexed by a macrocyclic ligand whose structure constrains the stereochemistry of the copper to be intermediate between tetrahedral and square-planar, providing a compromise between the requirements of the two oxidation

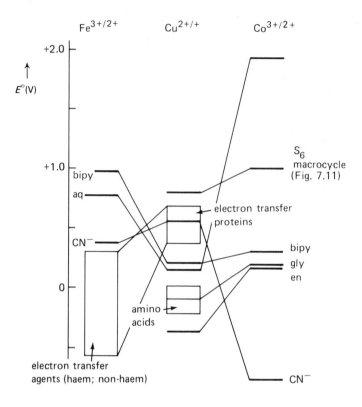

Fig. 7.12 — Comparison of redox potentials for copper(II)/(I) complex couples with those for analogous iron(III)/(II) and cobalt(III)/(II) couples (in all cases for the complexes containing the maximum number of ligands rather than for ternary aqua-ligand species).

states. Fig. 7.12 shows how the range of redox potentials for these biological copper(II)/(I) couples complements that of the haem and non-haem iron(III)/(II) electron transfer couples.

This section ends with a caution in relation to the use of redox potentials in discussions of ill-characterised biochemical, geochemical, or industrial systems. The often large effects of ligands on $E^\ominus$ values, for $M^{n+}/M^{(n-1)+}$ couples in general, mean that any temptation to use $E^\ominus(M^{n+}/M^{(n-1)+})$ in place of unavailable $E^\ominus$ $(ML_x^{n+}/ML_x^{(n-1)+})$ should be firmly resisted.

7.4   REDOX POTENTIALS AND STABILITY CONSTANTS

At this point it is useful to bring together this and the preceding chapter, through a cycle linking redox potentials for aqua-ions and complexes with stability constants for the latter (Fig. 7.13(a)). This cycle is often of practical utility. The redox potential for the aqua-complex couple is generally known. If the stability constants for the complexes of the $n+$ and $(n-1)+$ ions are both known, then the redox potential for the complex can be obtained. On the other hand it is often easier to measure this

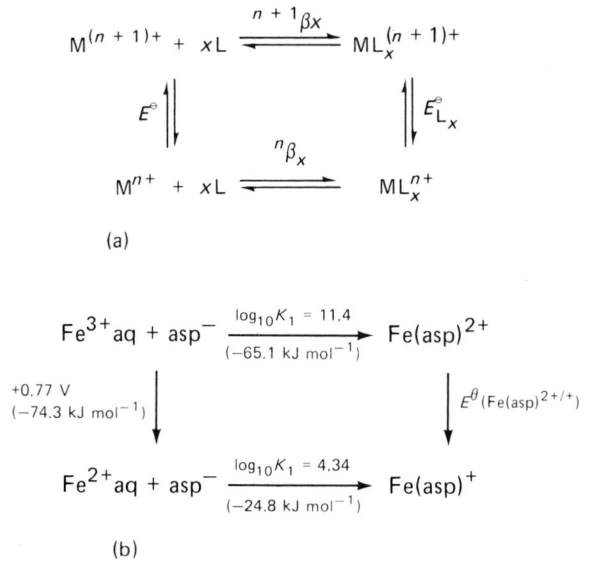

$$M^{(n+1)+} + xL \xrightleftharpoons{\;^{n+1}\beta_x\;} ML_x^{(n+1)+}$$

$$E^{\ominus} \big\updownarrow \qquad\qquad\qquad \big\updownarrow E_{L_x}^{\ominus}$$

$$M^{n+} + xL \xrightleftharpoons{\;^{n}\beta_x\;} ML_x^{n+}$$

(a)

$$Fe^{3+}aq + asp^{-} \xrightarrow[(-65.1 \text{ kJ mol}^{-1})]{\log_{10} K_1 = 11.4} Fe(asp)^{2+}$$

$$\begin{array}{c} +0.77 \text{ V} \\ (-74.3 \text{ kJ mol}^{-1}) \end{array} \Bigg\downarrow \qquad\qquad\qquad \Bigg\downarrow E^{\theta}(Fe(asp)^{2+/+})$$

$$Fe^{2+}aq + asp^{-} \xrightarrow[(-24.8 \text{ kJ mol}^{-1})]{\log_{10} K_1 = 4.34} Fe(asp)^{+}$$

(b)

From this cycle, the redox potential for the $Fe(asp)^{2+/+}$ couple can be estimated thus:

$$\Delta G^{\ominus} = -nFE^{\ominus} = -34.0 \text{ kJ mol}^{-1} \text{ (Hess's Law)}$$

{In fact this redox potential has been measured as +0.33 V. The slight difference between measured and estimated values arises from differences in experimental conditions (different ionic strengths: different acid media: etc.) between the experiments undertaken to measure the stability constants for the two complexes, and between these and the standard conditions for redox potentials.}

Fig. 7.13 — Cycles relating redox potentials and stability constants: (a) general and (b) specific application to iron(III)/(II)/aspartate.

redox potential and one of the stability constants, and use the cycle to obtain the other. The application of the cycle to the amino acid complexes of iron(II) and iron(III) included in Fig. 7.8 is detailed in Fig. 7.13(b).

8

Kinetics and thermodynamics

In chapters 4 to 7 we have dealt with several aspects of thermodynamics of ions in solution. In Chapters 9 to 12 we shall deal with kinetic and mechanistic aspects of the behaviour of ions in solution. This short chapter links these two themes, trying to show where there are correlations, and where it is inadvisable to make too direct a connection between kinetics and thermodynamics. The most important point to emphasise is that thermodynamic data apply to equilibrium (final) states, whereas kinetic data refer to the approach to final (equilibrium) states. There is not necessarily any direct link between the distance of a system from equilibrium and its rate of approach to the equilibrium state. Thus, for example, peroxodisulphate, $S_2O_8^{2-}$, is a very strong oxidant ($E^\ominus = +2.12\,\text{V}$), but is generally very slow to oxidise even strongly reducing substrates. The reason for this is the necessity to break the strong O–O bond *en route* to the sulphate ions which are the normal product. Both peroxodisulphate and permanganate ($E^\ominus = +1.51\,\text{V}$) are thermodynamically unstable with respect to reduction by water ($E^\ominus = +1.23\,\text{V}$), but aqueous solutions of their salts can be kept for extended periods with very little decomposition. On the other hand, redox reactions which involve simply the transfer of an electron, for example electron exchange between ferrocene and the ferrocinium cation (Fig. 8.1) or the oxidation of Fe^{2+} aq by Ce^{4+} aq, are often fast. Indeed such reactions often do exhibit kinetic–thermodynamic correlations (Chapter 11).

There is, of course, a direct connection between equilibrium constants and the rate constants for their forward and reverse reactions:

$$A + B \underset{k_b}{\overset{k_f}{\rightleftharpoons}} C + D \qquad K = \frac{k_f}{k_b}$$

The rate constant for approach to equilibrium is the sum of the rate constants for the forward and reverse reactions. There need be no correlation between K and k_f (or k_b) for a series of related reactions. An oft-quoted example of this is cyanide exchange at cyanocomplexes of transition metals. The available thermodynamic and

Table 8.1 — Thermodynamic and kinetic data relating to cyanide exchange at cyano-complexes of transition metals

Complex	$\log_{10}\beta_n$	Mean ΔH(M–CN) (kJ mol^{-1})	k(*CN$^-$ exchange) (s^{-1})	
			fast	slow
$[Mn(CN)_6]^{4-}$		-24	$>10^{-2}$	
$[V(CN)_6]^{4-}$		-33	$>10^{-2}$	
$[Co(CN)_5]^{3-}$	19	-43	$>10^{-2}$	
$[Cr(CN)_6]^{4-}$		-44	$>10^{-2}$	
$[Mn(CN)_6]^{3-}$			2×10^{-4}	
$[Ni(CN)_4]^{2-}$	31	-45	$>10^{-2}$	
$[Cr(CN)_6]^{3-}$				3×10^{-7}
$[Fe(CN)_6]^{4-}$	34	-60		$<10^{-6}$
$[Pt(CN)_4]^{2-}$	35		1.2×10^{-2}	
$[Pd(CN)_4]^{2-}$	42	-96	$>10^{-2}$	
$[Fe(CN)_6]^{3-}$	44	-49		$<10^{-6}$
$[Co(CN)_6]^{3-}$	64			$<10^{-6}$

kinetic data are set out in Table 8.1. In fact there is fast exchange at the complexes with lower stability constants and weaker metal–cyanide bonds and slow exchange at most of the complexes with high stabilities and strong metal–cyanide bonds — but the case of palladium(II) is a striking exception. Here the rate constant for cyanide exchange is very high, despite the high stability and exceptionally strong metal–cyanide bonding. The reason for this deviation from the general kinetic–thermodynamic trend must be the availability of a low energy associative path for this square-planar complex. Cyanide exchange with the octahedral complexes of cobalt(III), iron(III), and iron(II) presumably proceeds only by the more difficult dissociative mechanism. Cyanide exchange at platinum(II) also crops up in Table 8.2, which shows a series of ligand exchange rate constants for this metal centre. In these exchange reactions rate constants are actually *inversely* proportional to thermodynamic stabilities.

Sometimes equilibrium constants for a series of reactions are controlled by k_f, with k_b approximately constant, or vice versa. Rate constants for complex formation between alkali metal cations and crown ethers or cryptands show very little variation with the nature of the reactants or the solvent, but rate constants for dissociation correlate, over a range of more than ten orders of magnitude, with the respective stability constants (Fig. 8.1). Similarly, rate constants for complex formation between Ni^{2+}aq and simple uncharged ligands all fall within a narrow band (Table 8.3). Again the range of stability constants ($\log_{10}K_1$) observed is due almost entirely to the large variation in rate constants for dissociation. If, however, we consider the full range of complex formation reactions of Ni^{2+}aq, including ligands

Table 8.2 — Rate constants for ligand exchange (k_{ex}; in 0.05 M aqueous solution at 298.2 K) and respective stability constants (molar scale; 298.2 K) for a selection of platinum(II) complexes

Complex	k_{ex} (s^{-1})	$\log_{10}\beta_4$
$[Pt(CN)_4]^{2-}$	1.2×10^{-2}	35
$[PtI_4]^{2-}$	2.3×10^{-3}	25
$[PtBr_4]^{2-}$	1.5×10^{-3}	21
$[PtCl_4]^{2-}$	4.2×10^{-5}	16

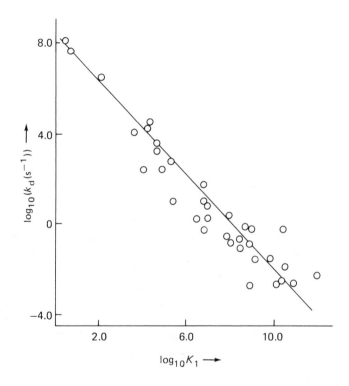

Fig. 8.1 — Correlation of logarithms of rate constants for dissociation of crown ether and cryptand complexes of alkali metal cations with logarithms of their respective stability constants (six solvents; 298.2 K).

of various charges, denticity, and other properties, then a graph such as that shown in Fig. 8.2 is obtained. The points marked ●, and the associated horizontal line, refer to the uncharged ligands mentioned in Table 8.3. The sloping line indicates the sort of

Table 8.3 — Formation rate constants $(k_f(M^{-1}s^{-1}))$ and stability constants K_1 (molar scale) for nickel(II) complexes of simple uncharged ligands in aqueous solution at 298.2 K

Ligand	$\log_{10}k_f$	$\log_{10}K_1$
bipy	3.2	6.8
histidine	3.4	
HF	3.5	
pyridine	3.6	1.7
phen	3.6	8.8
imidazole	3.7	3.3
ammonia	3.7	2.6

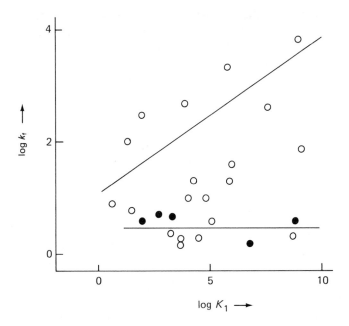

Fig. 8.2 — Relation between logarithms of rate constants for complex formation and logarithms of stability constants for nickel(II) complexes: ● simple uncharged ligands, ○ other ligands.

correlation that might be expected if rate constants, and stability constants, depended simply on ligand charge. Clearly the scatter exhibited by the data points in Fig. 8.2 demonstrates that there is no kinetic–thermodynamic correlation if we consider the whole range of complexing ligands. The reasons behind the scatter should become clearer after reading Chapter 10.

Before leaving the question of relations between kinetics and thermodynamics, it might be as well to clarify the meanings of the terms 'stable' and 'unstable', 'inert' and 'labile'. The latter pair of words apply to kinetics, and refer to species which undergo substitution or dissociation slowly and rapidly respectively. As stated above, there is no guarantee that rates correlate with thermodynamic properties, so it is inadvisable and potentially confusing to follow the careless practice sometimes encountered of using the terms 'stable' and 'unstable' in a kinetic context. 'Stable' and 'unstable' are strictly thermodynamic terms. Even when this is recognised, it is still sometimes necessary to state what the (in)stability is with respect to — dissociation of a complex into metal ion and ligands, hydrolysis by water, or oxidation by atmospheric oxygen, to cite the most common sources of instability in everyday chemical experience.

In the following three chapters we shall cover three important reactions of ions in solution. These will be solvent exchange, complex formation, and oxidation-reduction (redox) reactions. The first two of these apply to solvated metal ions, but redox reactions may involve cations, complexes, anions, and uncharged species. In a sense it would be logical to start with complex formation, following on from the thermodynamic treatment of stabilities of Chapter 6. However, the solvent exchange reaction is more fundamental — the simplest case of complex formation, in fact — so we shall start there (Chapter 9).

9

Kinetics and mechanisms: solvent exchange

9.1 INTRODUCTION

The simplest substitution reaction of a solvated ion in solution is that of exchange of solvent molecules between the ion solvation shell and bulk solvent. Knowledge of solvation shells is much more detailed for metal cations than for anions, and indeed rates of exchange are often within the range amenable to study for the former, rarely for the latter. Hence very much more is known about the dynamics of solvent exchange at metal ions than at anions. In this chapter we shall deal only with exchange at the former. The two basic questions concern mechanism — associative or dissociative — and reactivity. Fig. 9.1 gives an idea of the enormous range of rate constants known for water exchange at metal ions in aqueous solution. There is a range of over 16 orders of magnitude between Cu^{2+}aq and Cr^{2+}aq on the one hand, and Rh^{3+}aq on the other (and presumably water exchange at Ir^{3+}aq is yet slower). Even restricting attention to ions of the same charge and approximately the same radius, there is still a range of well over a million times along the 2+ cations of the first row of d-block metals. Variation of solvent at a given metal ion also gives a large range of rate constants (Table 9.1). We need to discuss mechanisms now, and then return to the question of reactivity later.

9.2 MECHANISMS

The reaction of solvent exchange is very simple. The reagent and product are identical, as are the ligand and solvent. The leaving and entering groups are identical; the Gibbs free energy change for the reaction is zero. Unfortunately simplicity brings problems when trying to establish mechanisms. Perhaps the most important means for mechanism diagnosis is normally the determination of the rate law for a reaction. In the case of solvent exchange we cannot vary the concentration of the solvent (without varying the medium), so lose this most important variable. It is possible to circumvent this difficulty to some extent by using small and varying

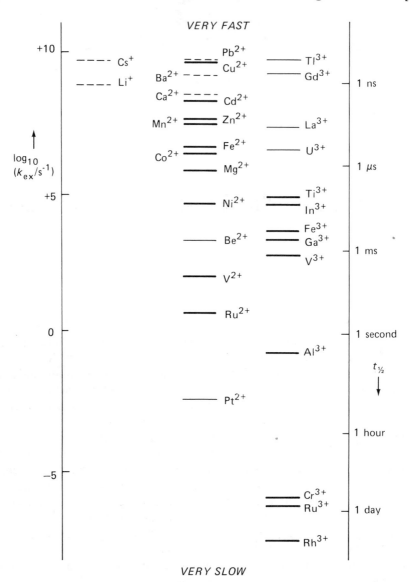

Fig. 9.1 — Rate constants for water exchange, and mean residence times for water molecules in primary hydration shells, for 2+ and 3+ metal ions, at 298.2 K. Octahedral species are indicated by thick lines (——), non-octahedral species by thin lines (——). Dashed lines (– – –) denote estimates derived from rate constants for complex formation.

concentrations of the exchanging solvent in an inert medium. Such weakly or non-coordinating solvents as acetone, nitromethane, and dichloromethane are popular inert co-solvents in such investigations, generally in their perdeutero-forms as NMR techniques are widely used in the determination of rate constants for solvent exchange. Several examples of this approach are given in Table 9.2. These illustrate several factors which may affect the mechanism:

Table 9.1 — Dependence of rate constants $(\log_{10}(k_{ex}/s^{-1})$ at 298.2 K) for solvent exchange on the nature of the solvent

	Ni^{2+}	Co^{2+}	Al^{3+}	Cr^{3+}
Liquid ammonia	5.0	6.9		-5.1
Water	4.5	6.3	-0.8	-6.3
Ethanol	4.0			
Dimethylformamide	3.8	5.5	-0.8	-7.3
Acetonitrile	3.7	5.3		
Dimethyl sulphoxide	3.7	5.2	-1.2	-7.5
Methanol	2.7	4.0		

Table 9.2 — Rate laws and mechanisms of solvent exchange at metal cations in mixed solvents consisting of a coordinating and a non-coordinating component

Cation	Solvation number	Solvents[a]		Order[b]	Mechanism
		Coordinating	Non-coordinating		
(1) Be^{2+}	4	TMU		0	dissoc.
	4	DMSO	$MeNO_2$	1	assoc.
	4	TMP		1	assoc.
(2) Zn^{2+}	4	TMTU		0	dissoc.
	4	HMPA	CD_2Cl_2	1	assoc.
	6	TMP		0	dissoc.
(3) Al^{3+}	6			0	dissoc.
Ga^{3+}	6	TMP	$MeNO_2$	0	dissoc.
In^{3+}	6			1	assoc.
(4) Ga^{3+}	6			0	dissoc.
Er^{3+}	8	DMF	$MeNO_2$	0	dissoc.
Tm^{3+}	8			0	dissoc.

[a] Solvents: TMU tetramethylurea.
  TMP trimethyl phosphate.
  TMTU tetramethylthiourea.
  DMSO dimethyl sulphoxide.
  $MeNO_2$ nitromethane.
  CD_2Cl_2 dichloromethane.
  HMPA hexamethylphosphoramide.
[b] Order with respect to the coordinating solvent.

(1) The entries for Be^{2+} show that associative attack is relatively easy at this tetrahedral centre. Expansion of coordination number to five is facile, except for weakly coordinating or bulky solvents such as tetramethylurea.

(2) The balance between coordinating strength and bulk is shown by the first two entries for Zn^{2+}. Both tetramethylthiourea and hexamethylphosphoramide are bulky molecules, restricting the coordination number of the zinc to four; only

hmpa is sufficiently strongly donating to force associative attack despite steric difficulty in forming a five-coordinate transition state. Trimethyl phosphate exchange at the octahedral Zn^{2+} is dissociative (cf. (3)).

(3) The sequence Al^{3+} to Ga^{3+} shows the mechanism of exchange changing from dissociative to associative as the size of the central metal ion increases and makes formation of a transition state of increased coordination number easier.

(4) DMF exchange at the lanthanide cations Er^{3+} and Tm^{3+} is dissociative, presumably thanks to the difficulty of increasing the coordination number of these cations from eight to nine. However, there are indications that at the left-hand end of the lanthanide series, the larger radii of, for example, La^{3+} and Ce^{3+} permit some associative character to DMF exchange. For the much smaller solvent water, the mechanism of solvent exchange seems to be associative in character even for the smaller Ln^{3+} cations towards the end of the series.

There can be both practical and interpretative difficulties in using this approach. The practical difficulty is that of miscibilities, particularly for water as coordinating solvent. Water is miscible in all proportions with a variety of polar and therefore potentially coordinating and competitive solvents, but tends to be almost immiscible with the majority of the incontrovertibly non-coordinating solvents, such as hexane, toluene, or chloroform. Interpretatively, it is not always possible to be sure that inter-component interactions within the mixed solvent are not having an effect on rate constants.

The determination of activation enthalpies and entropies, from the variation of rate constants for solvent exchange with temperature, provides a better approach to the problem of mechanism diagnosis. The qualitative idea that associative mechanisms tend to have lower activation enthalpies is sometimes of use, but the sign and magnitude of the activation entropy is a much better guide. The general distinction is that associative mechanisms have negative activation entropies, dissociative positive. In an associative reaction, two species come together in forming the transition state. Thereby translational freedom, and probably some rotational and vibrational freedom as well, are lost, so that $\Delta S^{\neq}$ is negative. In a dissociative process, lengthening and weakening of the bond to the leaving group gives increased freedom of movement, and $\Delta S^{\neq}$ can be expected to be positive. These deductions assume that there is no significant contribution from solvational changes during formation of the transition state. It is therefore unwise to diagnose mechanisms from $\Delta S^{\neq}$ for reactions which involve charge separation, augmentation, or cancellation. But for solvent exchange, where the leaving and entering molecules are uncharged, $\Delta S^{\neq}$ values provide a useful, though not in practice an entirely infallible, guide. Table 9.3 includes a selection of examples. Where direct comparison is possible, the conclusions in this table are consistent with those from rate law evidence (Table 9.2). Thus both Table 9.2 and Table 9.3 show a change of mechanism from dissociative for Al^{3+} and Ga^{3+} to associative for the larger In^{3+}.

The entries for Be^{2+} in Tables 9.2 and 9.3, and for Sc^{3+} in Table 9.3, show that mechanisms of solvent exchange can depend on the nature of the solvent (even at constant solvation number). For systems near the borderline, dissociative and associative mechanisms may operate in parallel. This is apparent from rate laws

Table 9.3 — Activation entropies as a guide to solvent exchange mechanisms

Cation	Solvent[a]	$\Delta S^{\pm}$ (J K^{-1} mol^{-1})	Mechanism
Be^{2+}	TMU	+16	dissociative
	DMSO	−32	associative
	TMP	−54	
Al^{3+}	water	+42	
	DMSO	+22	dissociative
	TMP	+37	
	DMF	+43	
Ga^{3+}	water	+30	
	DMSO	+ 4	dissociative
	DMF	+46	
In^{3+}	water	−96	associative
	TMP	−113	
Sc^{3+}	TMP	−126	associative
	DMA	−132	
	TMU	+48	dissociative
Tm^{3+}	DMF	+10	dissociative
Cr^{3+}	DMSO	−49	associative
	DMF	−42	
Fe^{3+}	water	−54	
	MeOH	−31	associative
	DMSO	−43	
Pd^{2+}	water	−24	associative

[a] Solvent abbreviations as Table 9.3, plus: DMA dimethylacetamide; DMF dimethylformamide.

determined for exchange of a coordinating solvent in an appropriate diluent. The form

$$\text{rate} = k_1[M^{n+}] + k_2[M^{n+}][\text{solv}]$$

indicating such concurrent paths has been established for several solvent exchange reactions of, for example, Be^{2+} and Sc^{3+}. Activation parameters for exchange of *NN*-dimethylformamide (DMF; HCONMe$_2$) and of *NN*-dimethylacetamide (DMA; MeCONMe$_2$) at Be^{2+} are given in Table 9.4. In each case the k_1 path has a positive $\Delta S^{\pm}$, the k_2 path a negative $\Delta S^{\pm}$, corresponding to dissociative and associative mechanisms respectively. Moreover the $\Delta H^{\pm}$ values for dissociative paths are higher than those for associative paths (see the start of the previous paragraph). It must be admitted that the situation is not always quite as clear-cut as in the examples chosen for illustration here.

Table 9.4 — Activation entropies and enthalpies for solvent exchange at Be^{2+} in systems obeying a two-term rate law

Coordinating solvent	Diluent	$\Delta S^{\ddagger}$ (J K^{-1} mol^{-1})		$\Delta H^{\ddagger}$ (kJ mol^{-1})	
		k_1 path	k_2 path	k_1 path	k_2 path
DMF	CD_3NO_2	+16	−32	84	58
DMA	CD_3CN	+8	−27	78	68

Discussion of mechanisms of solvent exchange at transition metal ions appears below, after the introduction of activation volumes as a mechanistic criterion. However, it is convenient to comment at this stage on the activation entropy for water exchange at Pd^{2+}aq. The markedly negative value (Table 9.3) is consistent with the expectation of associative substitution at this square-planar d^8 centre.

Activation volumes, derived from the pressure dependence of rate constants $(\partial \ln k / \partial P = -\Delta V^{\ddagger}/RT)$, are perhaps the most attractive way of assigning solvent exchange mechanisms, although technically they are harder to obtain than activation entropies. It is much easier to measure rate constants over a small range of temperatures at ambient pressure than to measure rate constants at pressures up to 1 kbar or more in a well-thermostatted apparatus. Developments in apparatus and techniques over the past two decades have resulted in a marked increase in the number of $\Delta V^{\ddagger}$ values available. This increase is shown, for inorganic reactions, in Fig. 9.2. A significant contribution to this trend has been made by solvent exchange studies, in response to the problems of mechanism assignment set out earlier in this section. In this connection, the most important development has been the construction of apparatus for running NMR spectra at high pressures (up to about 2 kbar). Fig. 9.3 shows the effect of pressure on the ^{1}H NMR spectrum of an Al^{3+}-trimethyl phosphate-nitromethane solution over the range 0 to 200 MPa (0 to 2 kbar; at 341 K). Increasing the pressure slows down solvent exchange so that the relatively broad signal at atmospheric pressure due to relatively fast bulk $\rightleftharpoons$ coordinated interchange (see Chapter 2) becomes sharper as the average residence time of TMP solvent molecules in the primary coordination sphere becomes significantly longer than the time constant corresponding to the NMR frequency.

Activation volumes offer several advantages over activation entropies. In the first place, volumes are more readily visualised (Fig. 9.4) both in respect of their sign and in relation to the sizes of the reactant(s) and product(s) of reaction. Secondly, and more importantly, they are often more accurate and reliable. $\Delta V^{\ddagger}$ comes from the *slope* of a log k vs. P plot, whereas $\Delta S^{\ddagger}$ comes from the *intercept*, usually far distant from the experimental points, of the log k vs. $1/T$ plot. In the particular case of deriving rate constants from NMR spectra taken over a range of temperatures, there can be problems in the extraction of rate constants for the chemical exchange process. There are often other dynamic processes which restrict the temperature range for an Arrhenius plot, and may interfere with the accuracy of the derived rate

Activation volumes

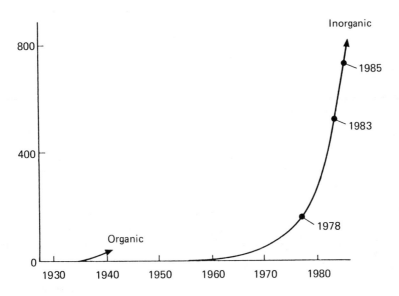

Fig. 9.2 — The increasing availability of activation volume data for organic and inorganic reactions.

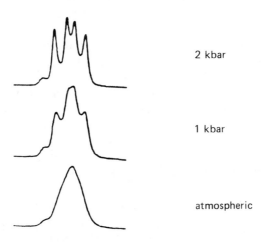

Fig. 9.3 — The effect of increasing pressure on the 60 MHz ^{1}H NMR spectrum of a trimethyl phosphate–nitromethane (CD_3ND_2) solution containing Al^{3+}, at 341.3 K.

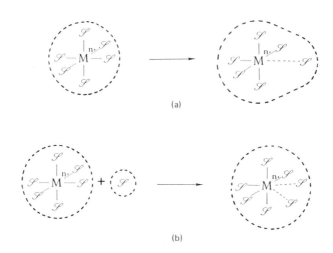

Fig. 9.4 — Volume changes on transition state formation for (a) dissociative and (b) associative solvent exchange.

constants at the ends of the appropriate temperature range. These problems are too complicated to discuss here†, but their consequences can be all too apparent. Thus, for example, a succession of authors have over the years reported $\Delta S^{\ddagger}$ values fairly evenly spread over the range -80 to $+10$ J K^{-1} mol^{-1} for exchange of dimethyl sulphoxide at Ni^{2+}, and -37 to $+50$ J K^{-1} mol^{-1} for exchange of acetonitrile at this same cation. The actual rate constants determined for acetonitrile exchange at Ni^{2+} at 298 K cover a range of only twofold. In certain cases it is possible to combine results from standard NMR techniques with those from stopped-flow Fourier transform NMR experiments to give data over a temperature range of the order of 100°C, but this can only be done for a few systems (and even then it has to be tacitly assumed that the Arrhenius plot is linear, i.e. $\Delta C_p^{\ddagger}$ is negligible — which is probably true here). Table 9.5 gives some examples of solvent exchange reactions where activation entropies and volumes, and indeed rate laws in some cases, lead to the same mechanistic conclusions. It should be added that activation volumes, like activation entropies, are in general liable to significant solvation contributions, but not in solvent exchange where entering and leaving molecules are uncharged.

Perhaps the most important application of $\Delta V^{\ddagger}$ determinations in relation to solvent exchange has been the demonstration of a smooth change in solvent exchange mechanism across the first row of the d-block transition elements. Both for 2+ and for 3+ ions, there is a steady trend from associative to dissociative character as one goes from left to right, from d^0 to d^{10} (Table 9.6). The changeover from

---

† For amplification and explanation of these difficulties, see pp. 44–45, 73–74, and 404–408 of *Inorganic High Pressure Chemistry: Kinetics and Mechanisms*, R. van Eldik (ed.), Elsevier (1986). The particular case of acetonitrile exchange at nickel(II) is dealt with in K. E. Newman, F. K. Meyer and A. E. Merbach, *J. Am. Chem. Soc.*, **101**, 1470 (1979).

Table 9.5 — Activation volumes and entropies for solvent exchange

Cation	Solvent	$\Delta V^{\ddagger}$ (cm³ mol⁻¹)	$\Delta S^{\ddagger}$ (J K⁻¹ mol⁻¹)
Al^{3+}	DMSO	+16	+22
Ga^{3+}	water	+5	+30
Co^{2+}	water	+6	+37
V^{3+}	water	−9	−28
Cr^{3+}	DMSO	−11	−64

Table 9.6 — Activation volumes (cm³ mol⁻¹) for solvent exchange at metal ions

	d^0	d^1	d^2	d^3	d^4	d^5	d^6	d^7	d^8	d^9	d^{10}
				V^{2+}		Mn^{2+}	Fe^{2+}	Co^{2+}	Ni^{2+}	Cu^{2+}	
Water				−4.1		−5.4	+3.8	+6.1	+7.2		
MeOH						−5.0	+0.4	+8.9	+11.4	+8.3	
MeCN						−7.0	+3.0	+7.7	+9.6		
DMF							+8.5	+6.7	+9.1		

	Sc^{3+}	Ti^{3+}	V^{3+}	Cr^{3+}	Fe^{3+}	Ga^{3+}
Water		−12	−8.9	−9.3	−5.4	+5
DMF				−6.3	−0.9	+7.9
DMSO			−11		−3.1	+13.1
TMP	−20.7					+20.7

associative			dissociative

associative to dissociative character is clearly apparent in $\log(k_p/k_0)$ against pressure plots. Fig. 9.5 shows how water exchange rate constants decrease with increasing pressure for V^{2+}aq and Mn^{2+}aq, but increase with increasing pressure for the transition metal 2+ cations with more d electrons (Fe^{2+}, Co^{2+}, Ni^{2+}). There is a similar contrast for the 3+ cations between the accelerating effect of increased pressure on associative water exchange at the 3+ transition metal cations included in this figure, and the opposite for Ga^{3+}aq. The data in Fig. 9.5 and in Table 9.6 for the oxidation state III cations refer specifically to M^{3+}aq. It is interesting to note that whereas water exchange at Fe^{3+}aq is associative in character, water exchange at $Fe(OH)^{2+}$aq is dissociative; $\Delta V^{\ddagger}$ values are −5.4 and +7.0 cm³ mol⁻¹ respectively. Similarly, $\Delta V^{\ddagger}$ values of −9.6 and +2.7 cm³ mol⁻¹ for water exchange at Cr^{3+}aq and at $Cr(OH)^{2+}$aq indicate predominantly associative and dissociative mechanisms respectively here too. However, $\Delta V^{\ddagger}$ values of +5.0 and +6.2 cm³ mol⁻¹ for water exchange at Ga^{3+}aq and at $Ga(OH)^{2+}$aq suggest dissociative activation in both cases this time.

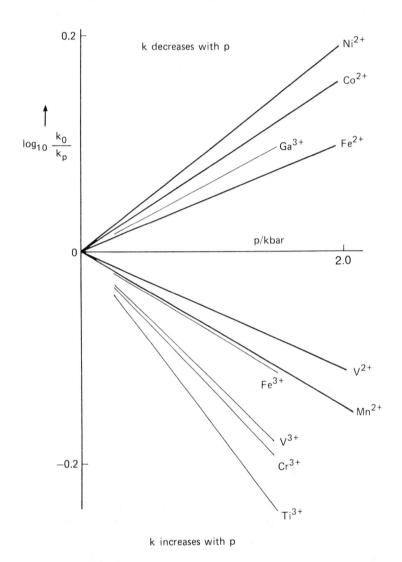

Fig. 9.5 — Pressure effects on rate constants for water exchange at 2+ and 3+ metal ions (first row).

We have earlier shown how rate laws (Table 9.2) and activation entropies (Table 9.3) both indicate changeovers from dissociative to associative activation on descending Groups in the Periodic Table. Activation volumes can provide similar information. Table 9.7 shows this for Group III sp-block elements, where $\Delta V^{\ddagger}$ and $\Delta S^{\ddagger}$ give consistent indications. Table 9.8 gives an example of dissociative to associative changeover signalled by $\Delta V^{\ddagger}$ for ternary aqua-complexes of a triad of

Table 9.7 — Activation volumes and activation entropies as indicators of solvent exchange mechanisms

Cation	Solvent	$\Delta V^{\pm}$ (cm^3 mol^{-1})	$\Delta S^{\pm}$ (J K^{-1} mol^{-1})	Mechanism
Al^{3+}	TMP^a	+23	+38	dissociative
Ga^{3+}	TMP	+21	+27	
In^{3+}	TMP	−21	−118	associative
Sc^{3+}	TMP	−21	$ca.$ −100	

^{a}TMP = trimethyl phosphate.

Table 9.8 — Activation volumes (cm^3 mol^{-1}) for solvent exchange at complexes $[M(NH_3)_5(solv)]^{3+}$

	d^3		d^6	
Aqueous	$[Cr(NH_3)_5(OH_2)]^{3+}$	− 5.8	$[Co(NH_3)_5(OH_2)]^{3+}$	+ 1.2
			$[Rh(NH_3)_5(OH_2)]^{3+}$	− 4.1
			$[Ir(NH_3)_5(OH_2)]^{3+}$	− 3.2
Non-aqueous	$[Cr(NH_3)_5(DMSO)]^{3+}$	− 3.2	$[Co(NH_3)_5(DMSO)]^{3+}$	+ 2.0
	$[Cr(NH_3)_5(DMF)]^{3+}$	− 7.4	$[Co(NH_3)_5(DMF)]^{3+}$	+ 2.6

transition metals. This table also provides evidence for the expected change from associative to dissociative activation for solvent exchange on going from the left-hand transition metal cation Cr^{3+} to Co^{3+}. Such evidence is only available from ternary complexes due to the instability of the majority of solventocobalt(III) cations, including of course Co^{3+}aq (though Co^{3+} in liquid ammonia could in principle be examined kinetically over a range of pressures). The study of the ternary complexes of the type included in Table 9.8 fills an important gap in the data presented in Table 9.6.

Finally it should be said that the activation volumes for water exchange at Pd^{2+}aq and Pt^{2+}aq, which are −2 and −5 cm^3 mol^{-1} respectively, are consistent with the associative activation expected for these square-planar cations. Here again there is agreement with the deductions from activation entropies (see Table 9.3).

9.3 REACTIVITIES

Reactivity trends across and down the Periodic Table are complicated by changes in mechanism as described in the previous section. However, it is possible to discern some general trends. Thus in the sp-block, there is evidently a considerable effect of cation charge and radius (Fig. 9.6), which dominates over changes in mechanism and over such subtle effects as 'hard' and 'soft' character.

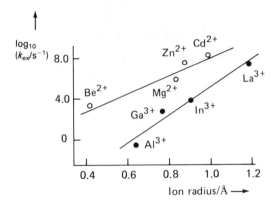

Fig. 9.6 — The dependence of rate constants for water exchange on ionic radius for selected 2+ and 3+ metal ions.

The limited data for f-block cations (lanthanide(III) and actinide(IV)) cations can be accommodated within this electrostatic framework, but reactivities for d-block cations are controlled to a very great extent by Crystal Field effects. Rate constants for water exchange at the first row M^{2+}aq ions range from 10^2 to $10^9 s^{-1}$ (cf. Fig. 9.1) despite constancy of charge and very small range of ionic radii (0.84 to 0.96 Å). For M^{2+}aq ions of sp-block metals, rate constants for water exchange are of the order of 10^5 to $10^8 s^{-1}$, so Crystal Field effects can reduce reactivity by several orders of magnitude. The variation of water exchange rates across the first row of the d-block M^{2+}aq ions is shown in Fig. 9.7. Crystal Field effects are abundantly clear,

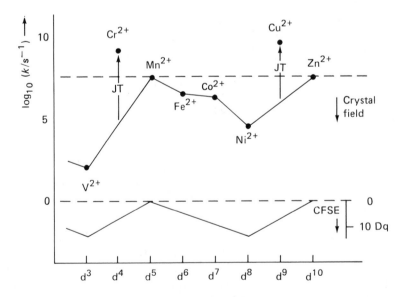

Fig. 9.7 — The dependence of rate constants for water exchange on Crystal Field and Jahn–Teller effects, i.e. on d electron configuration, for first row transition metal 2+ ions.

and are not clouded by the change in mechanism across this series of aqua-ions (section 8.2). Rate constants are dramatically higher than expected from Crystal Field forecasts for Cr^{2+}aq and Cu^{2+}aq. For these d^4 and d^9 ions, marked Jahn–Teller distortions from octahedral geometry mean that two of the six water molecules in the primary hydration shell are rather distant. These longer bonds are weaker, and thus water exchange can take place more readily and more rapidly.

There is rather little information on solvent exchange reactivity trends on going down Groups in the Periodic Table. In Group III, rate constants increase steadily and greatly going from aluminium down to thallium (Fig. 9.1). This trend presumably reflects decreasing metal–solvent bond strengths as well as progressively easier attack by the associative route. In the d-block, water exchange at Mo^{3+}aq is much faster than at Cr^{3+}aq, presumably again attributable to increasingly easy associative attack. But by the time the cobalt–rhodium–iridium triad is reached, solvent exchange rates *decrease* as we descend the Periodic Table. Now reactivity must be dominated by bond strengths, with the greater Crystal Field contributions to bonding exerting an overriding effect for rhodium and, even more, iridium.

10

Kinetics and mechanisms: complex formation

10.1 BACKGROUND

The formation of a metal complex from a solvated metal ion and a ligand is, like solvent exchange, a special case of substitution (Fig. 10.1). It is a special case which is

---

SUBSTITUTION: GENERAL

$$ML_5L' + L'' \rightarrow ML_5L'' + L'$$

e.g. $[Fe(CN)_5(NH_3)]^{3-} + py \rightarrow [Fe(CN)_5(py)]^{3-} + NH_3$

SUBSTITUTION: SPECIFIC

Solvent exchange

$$MS_6^{n+} + {}^*S \rightarrow MS_5{}^*S^{n+} + S$$

e.g. $[Al(OH_2)_6]^{3+} + {}^*OH_2 \rightarrow [Al(OH_2)_5({}^*OH_2)]^{3+} + OH_2$

Complex formation

$$MS_6^{n+} + L \rightarrow MS_5L^{n+} + S$$

e.g. $[Ni(OH_2)_6]^{2+} + Br^- \rightarrow [Ni(OH_2)_5Br]^+ + OH_2$

Aquation or solvolysis

$$ML_5L' + S \rightarrow ML_5S + L'$$

e.g. $[Co(NH_3)_5Cl]^{2+} + H_2O \rightarrow [Co(NH_3)_5(OH_2)]^{2+} + Cl^-$

Ligand exchange

$$ML_6^{n+} + {}^*L \rightarrow ML_5{}^*L^{n+} + L$$

e.g. $[Fe(CN)_6]^{4-} + {}^*CN^- \rightarrow [Fe(CN)_5({}^*CN)]^{4-} + CN^-$

Fig. 10.1 — Types of substitution reactions at complexes.

not only of considerable interest in its own right, but also of importance in a variety of applications, for example in analysis (solvent extraction in particular) and in biological chemistry. On grounds of history, chemistry, and convenience, the treatment in this chapter will start with details of formation reactions of nickel(II). These reactions have played a key role in the development of this subject, as they can be carried out conveniently and quickly in a stopped-flow apparatus (Fig. 10.2). No other metal ion behaves so well chemically and reacts in such a convenient time-scale. Having established the general kinetic and mechanistic picture for complex formation reactions at solvated Ni^{2+}, the discussion will be extended to other metal ions and from simple monodentate ligands to chelating and macrocyclic ligands.

In the early days of the study of kinetics of complex formation, a picture began to emerge of a rate law which was first-order in metal ion concentration and first-order in incoming ligand concentration. The simplest rationalisation was in terms of an associative mechanism involving bimolecular reaction between metal ion and ligand. However, there were several unattractive features of such an interpretation. Firstly, rate constants showed much less dependence on the nature of the incoming ligand than would be expected, and activation enthalpies and entropies also covered a much smaller range than expected. Also there were then strong indications from other sources that the majority of octahedral metal(II) centres underwent substitution by dissociative mechanisms.

The experiments and speculations of Eigen, Tamm, and Wilkins led to a simple hypothesis which accommodated the experimental kinetic results for the whole range of metal ions, could accommodate dissociative or associative activation, and could be extended and modified to cope with polydentate and other complicated ligands. What has become known as the Eigen–Wilkins mechanism for complex formation is set out in Fig. 10.3, and its application to reactions between solvated Ni^{2+} and simple monodentate ligands discussed in the following section.

10.2   THE EIGEN–WILKINS MECHANISM

Table 10.1 presents a series of experimentally determined rate constants, k_f, for formation reactions of $Ni^{2+}aq$. This table also includes equilibrium constants, K_{os}, for association between $Ni^{2+}aq$ and each ligand. Those K_{os} values are calculated values, based on electrostatic considerations; these formation reactions take place much too rapidly for direct experimental determination of K_{os}. In the final column are given derived values for k_i, the interchange rate constant characterising replacement of a coordinated water molecule by the incoming ligand. The near constancy of the k_i values strongly suggests a common rate-determining step, i.e. that the interchange process is dissociative in character. It is interesting to compare these k_i values with the rate constant for water exchange at $Ni^{2+}aq$ ($3 \times 10^4 \, s^{-1}$); as should be the case for common rate-determining water loss from the Ni^{2+} in water exchange as in complex formation, k_i and k_{ex} values are indeed pretty well equal. The analysis in Table 10.1 thus shows that the experimental demonstration of a first-order term in incoming ligand concentration in the rate law can be interpreted as well through pre-association of the reactants followed by a dissociative rate-limiting step as through a one-step associative attack at $M^{n+}aq$. Differences in rate constants k_f between different ligands are generally a result of different charges, which affect reactivity through the size of the pre-association equilibrium constant K_{os} rather than through k_i.

Similar conclusions as to the operation of a common rate-determining interchange step in reactions of $Ni^{2+}aq$ with simple ligands can also be reached from inspection of activation enthalpies and volumes (Table 10.2). In particular, the

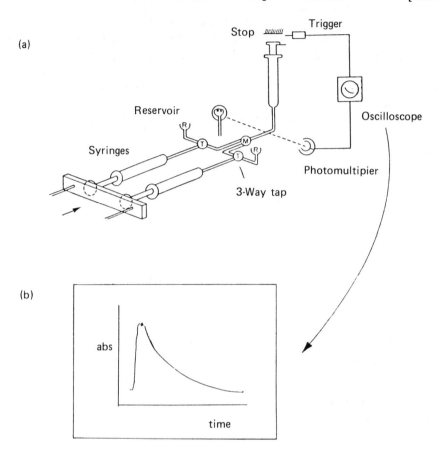

Fig. 10.2 — Stopped-flow kinetics: (a) schematic diagram of the apparatus; (b) absorbance-time trace as seen on the oscilloscope.

positive activation volumes for interchange strongly support dissociative character. Overall reaction volume changes have also been included in Table 10.2; volume profiles for several formation reactions of Ni^{2+}aq are given in Fig. 10.4. This figure includes the electrostriction contribution, $\Delta V_{os}^{\ominus}$ (estimated), separated from the volume of activation for the interchange step. $\Delta V_i^{\ddagger}$ values are approximately equal, and indeed are close to the activation volume for water exchange at Ni^{2+}aq, included in Fig. 10.4. $\Delta V_i^{\ddagger}$ values are also practically independent of the partial molar volume of the incoming ligand. The evidence is consistently in favour of dissociative interchange.

An analogous analysis of complex formation kinetics in terms of the Eigen–Wilkins mechanism for Mg^{2+}aq is given in Table 10.3. This table includes some ligands of biological relevance which, of course, conform to the pattern followed by the simple inorganic ligands. As for Ni^{2+}aq, the interchange step is seen to be dissociative in character. Similar analyses of complex formation rate constants for several other aqua-metal(II) cations, such as Co^{2+}aq, Cu^{2+}aq, and Zn^{2+}aq, also

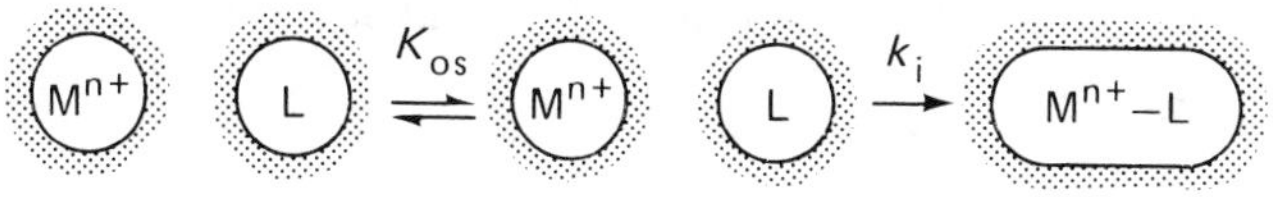

$$\mathrm{M(OH_2)}_x^{n+} + \mathrm{L} \xrightleftharpoons{K_{os}} \mathrm{M(OH_2)}_x^{n+}, \mathrm{L} \xrightarrow{k_i} \mathrm{M(OH_2)}_{x-1}\, L^{n+} + \mathrm{H_2O}.$$

General rate law:

$$+\frac{d[\mathrm{ML}^{n+}]}{dt} = \frac{K_{os}k_i[\mathrm{M}^{n+}][\mathrm{L}]}{1 + K_{os}[\mathrm{L}]}$$

Under usual experimental conditions, $[\mathrm{M}^{n+}] \gg [\mathrm{L}]$:

$$+\frac{d[\mathrm{ML}^{n+}]}{dt} = K_{os}k_i[\mathrm{M}^{n+}][\mathrm{L}]$$

Whence:

$$k_f = K_{os}k_i$$

Fig. 10.3 — Complex formation: mechanism, equations, and kinetics.

suggest dissociative interchange, though for some of these aqua-metal(II) ions data are rather sparse.

In Chapter 9 evidence was presented of a changeover of solvent exchange mechanism from dissociative to associative on going towards the left of the d-block series of transition metal $2+$ or $3+$ ions. This changeover was proposed on the basis of activation volume data. There are as yet relatively few activation volumes for complex formation reactions, but the values collected together in Table 10.4 suggest that the interchange for formation is predominantly associative for first row d-block aqua-metal(II) ions towards the left-hand side, viz. V^{2+} aq and Mn^{2+} aq.

The preceding paragraphs have dealt with aqueous solutions, but the available kinetic evidence indicates that the same pattern applies to the formation reactions of metal(II) cations in non-aqueous media. Thus, for example, the activation volumes reported in Table 10.5 indicate dissociative interchange of Fe^{2+}, Co^{2+}, and Ni^{2+} in NN-dimethylformamide as in water. There are indications that for the borderline case of Mn^{2+}, the mechanism of the interchange step may be influenced by the nature of the solvent and the incoming ligand.

If kinetic data for complex formation from several trivalent d-block metal cations are examined, it is found that now k_i depends strongly on the nature of the incoming group. The wide ranges of k_f values in Table 10.6 illustrates this, since as all the ligands have charge $1-$, K_{os} values are all approximately the same and the range in k_i will be about the same as the range in k_f. It is clear that the interchange step is associative for these $3+$ cations, with significant bonding to the incoming ligand in the transition state. As in the case of solvent exchange (Chapter 9), formation of complexes from Fe^{3+} aq and from $Fe(OH)^{2+}$ aq takes place by different mechanisms, predominantly associative and dissociative respectively. Again as for solvent

Table 10.1 — Rate constants and pre-association constants (defined in the text and in Fig. 10.3) for formation of complexes from Ni^{2+} aq, in aqueous solution at 298.2 K

Ligand	Measured $10^{-3}k_f(M^{-1}s^{-1})$	Estimated K_{os} (molar scale)	Derived $10^{-3}k_i(s^{-1})$
N-Methylimidazole$^+$	0.23	0.02	12
Imidazole H$^+$	0.3	0.02	15
Ammonia	5	0.15	33
Hydrogen fluoride	3	0.15	20
Imidazole	2.8–6.4	0.15	19–43
1,10-Phenanthroline	4.1	0.15	26
Diglycine	21	0.17	12
Fluoride$^-$	8	1	8
Acetate$^-$	100	3	30
Glycinate$^-$	20	2	10
Oxalate H$^-$	5	2	3
Oxalate^{2-}	75	13	6
Malonate^{2-}	450	95	5
Methylphosphate^{2-}	290	40[a]	7
Pyrophosphate^{3-}	2100	88	24
Tripolyphosphate^{4-}	6800	570	12
Cf. Water exchange			30

[a] In this favourable case K_{os} was derived from the kinetic results.

exchange, mechanisms for complex formation change from dissociative to associative on descending Group III. Activation volumes for formation from Al^{3+} aq and from Ga^{3+} aq are generally positive, from In^{3+} aq negative, with a parallel trend for these cations in dimethyl sulphoxide.

A general impression of relative reactivities of a range of metal cations with respect to complex formation is given in Fig. 10.5. In some cases, the uncertainty in the points is even greater than suggested in this figure. For several cations, kinetic data are very sparse or inconsistent, while for the cations which react by an associative mechanism, a range of rate constants will apply. Despite these uncertainties, a clear general picture emerges. For *sp*-block cations, the customary dependence on cation radius (its reciprocal in Fig. 10.5) and charge is apparent. However, the points do not all fall within one correlation band. The 1 + , 2 + , and 3 + cations give different plots; moreover the 'hard' and 'soft' sub-groups within Group II of the Periodic Table give separate though parallel correlation lines.

The sequence of reactivities for the first row transition elements M^{2+} aq cations is indicated by the thin line in Fig. 10.5. The span from top (Cu^{2+}) to bottom (V^{2+}) emphasises again the enormous consequences of crystal field effects and of the

Table 10.2 — Activation enthalpies and volumes, and overall volumes of reaction, for formation of complexes from Ni^{2+}aq, at 298.2 K

Ligand	$\Delta H_i^{\ddagger}$ (kJ mol^{-1})	$\Delta V_{os}{}^{a}$ (cm^3 mol^{-1})	$\Delta V_i^{\ddagger b}$ (cm^3 mol^{-1})	$\Delta V^{\ominus}$ (cm^3 mol^{-1})
Succinate^{2-}		+7	+6	+11
Malonate^{2-}	46	+7	+8	
Oxalate^{2-}	46 or 59			
Oxalate H^{-}	47 or 59			
Glycolate^{-}		+3	+11	+17
Glycinate^{-}	46	+3	+8	+2
Murexide^{-c}		+3	+9	+23
Ammonia	42	0	+7	−2
iso-Quinoline		0	+7	−2
pada^{d}	57	0	+8	+1

[a]Estimated.
[b]Derived from measured $\Delta V_i^{\ddagger}$ using given ΔV_{os}.
[c]Murexide^{-} =

[d]pada =

Jahn–Teller effect. The eccentric nature of this plot is a result of the fact that crystal field effects are reflected in ionic radii as well as reactivities, and that the Jahn–Teller effect is averaged in estimating effective ionic radii from crystal structures but leads to maximum reactivity in complex formation as in solvent exchange (see Chapter 9).

10.3 CHELATE FORMATION

The preceding discussion was developed mainly for monodentate ligands, with the incoming ligand simply replacing one outgoing solvent molecule. Although the same principles apply to reactions of solvento-cations with bidentate and polydentate ligands, such ligands do introduce a new feature, in the need to include chelate ring closure in the overall mechanism. There is also the problem of displacing several solvating molecules simultaneously if the incoming ligand is a rigid macrocycle.

Let us consider first the replacement of two solvating molecules by one molecule of a bidentate ligand. The elementary steps in such a reaction are shown in Fig. 10.6. The first step is identical with the first step in the Eigen–Wilkins mechanism for reaction with a monodentate ligand, in which pre-association is followed by rate-limiting interchange. In this interchange one end of the bidentate ligand becomes bonded to the metal ion, giving a transient intermediate. It is the next step, in which

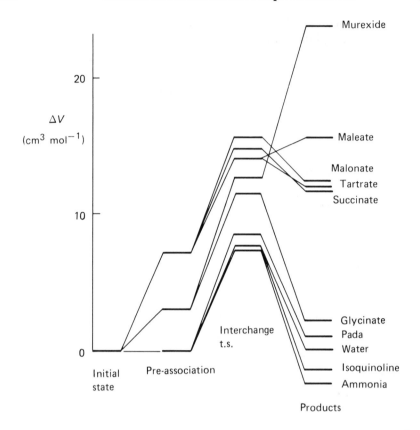

Fig. 10.4 — Volume profiles for complex formation from, and water exchange at, Ni^{2+} aq in aqueous solution.

the chelate ring closes with the displacement of a second solvating molecule from the metal ion that is of prime interest here. The overall kinetic pattern is controlled by the balance between the rate of solvent loss and the ease of chelate ring closure, which balance varies with the nature of the metal ion. For a cation such as Ni^{2+} which undergoes relatively slow exchange and formation reactions, loss of the second solvent molecule is usually rapidly followed by the second donor atom of the bidentate ligand bonding to the metal, taking advantage of the temporarily vacant site to close its chelate ring. The overall rate of formation of the bidentate complex is then similar to that for formation of an analogous monodentate ligand complex.

If the time-scale for solvent exchange is somewhat shorter, for example for Co^{2+} aq, or if there is some difficulty attendant on ring closure, as for example in the closure of a large chelate ring, then there may be competition between returning or replacing solvent molecules and the free end of the bidentate ligand for the vacant site left by the outgoing solvating molecule (see the inset in Fig. 10.6). As the competition favours solvent return more and more, so the rate of formation of the bidentate complex final product decreases. This is illustrated in Table 10.7 for the formation of complexes with different size chelate rings from Co^{2+} aq. It has been

Table 10.3 — Rate constants and preassociation constants (defined in the text) and activation enthalpies for formation of complexes from Mg^{2+}aq in aqueous solution at 298.2 K

Ligand	Measured $10^{-5}k_f(M^{-1}s^{-1})$	Estimated K_{os} (molar	Derived $10^{-5}k_i(s^{-1})$	$\Delta H^{\ddagger}$ (kJ mol^{-1})
Oxine[a]	0.13	0.2	0.7	40
Oxinate$^-$ [b]	6.0	2.1	2.9	51
Fluoride$^-$	0.55	1.6	0.4	
5-Nitrosalicylate$^-$	7.1	2	3.6	49
Bicarbonate$^-$	5.0	0.9	5.6	
Carbonate^{2-}	0.15	3.5	0.4	
PyrophosphateH$_2^{2-}$	5.4	13	0.4	
ADPH^{2-} [c]	10	9	1.1	
ATPH^{3-} [d]	30	30	1.0	
ATP^{4-} [d]	130	120	1.1	47
Cf. water exchange			1.0	43

[a]Oxine =   [b]Oxinate =

[c]ADP = adenosine diphosphate.
[d]ATP = adenosine triphosphate.

Table 10.4 — Activation volumes, $\Delta V_i^{\ddagger}$ (cm^3 mol^{-1}), for the interchange step in complex formation from first-row d-block aqua-metal cations

	V^{2+}	Mn^{2+}	Fe^{2+}	Co^{2+}	Ni^{2+} [a]	Cu^{2+}	Zn^{2+}
NCS$^-$	-5						
gly$^-$					$+8$	$+9$	$+4$
NH$_3$				$+5$	$+7$		
bipy		-1		$+6$	$+5$		$+4$
terpy		-3	$+4$	$+4$	$+6$		

[a]See also Table 10.2; $\Delta V_i^{\ddagger}$ for Ni^{2+} aq $+ L = +6$ to $+11$ cm^3 mol^{-1} for several other ligands L.

estimated that the rate constant for the ring closure step in the reaction of Co^{2+}aq with glycolate, giving a five membered chelate ring, is $1.2 \times 10^6\,s^{-1}$, whereas the rate constant for closure of the six-membered chelate ring from malonate is appreciably less, at $3.2 \times 10^5\,s^{-1}$.

Table 10.5— Activation volumes, $\Delta V_i^{\ddagger}$ ($cm^3\,mol^{-1}$), for complex formation from first-row d-block metal(II) cations in NN-dimethylformamide solution

	Fe^{2+}	Co^{2+}	Ni^{2+}
NCS$^-$			+9
Et$_2$dtc$^-$	+12	+12	+9
pada	+8	+10	+9
iso-quinoline			+9

Table 10.6 — Rate constants for formation of complexes from aqua-metal(III) cations of d-block elements, in aqueous solution at 298.2 K

	Ti^{3+} $10^{-3}k_f$ ($M^{-1}\,s^{-1}$)	V^{3+} k_f ($M^{-1}\,s^{-1}$)	Mo^{3+} $10^{-3}k_f$ ($M^{-1}\,s^{-1}$)
4-aminosalicylate$^-$		7000	
acetate$^-$	1800		
salicylate$^-$		1400	
oxalateH$^-$	390	1300	
3CNacac$^-$	160		
Cl$_2$CHCO$_2^-$	110		
N$_3^-$		(900)	
NCS$^-$	8	110	270
Br$^-$		10	
Cl$^-$		3	4.6
Cf. water exchange	100	180	

When solvent exchange is much faster, as for example with Mn^{2+}aq and Cu^{2+}aq, then solvent return may become very fast relative to ring closure. In such cases the relatively sluggish motion of the free end of the ligand through the medium means that its potential donor atom is very unlikely to be in exactly the right position to take advantage of the vacant site on the metal ion during the very short time before an adjacent molecule of solvent fills the space. Rate constants for formation of bidentate complex final products are now very much slower than rates of formation of analogous monodentate ligand complexes, and are markedly dependent on the geometry and structure of the ligand (Table 10.7 again). This mechanism, in which

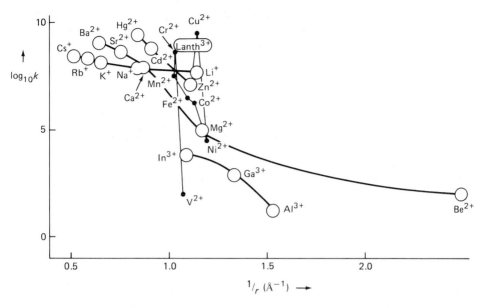

Fig. 10.5 — Relation of formation rate constants to ionic radii.

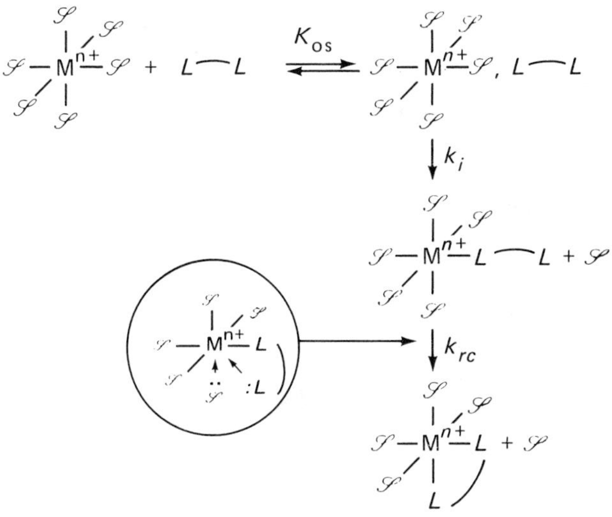

Fig. 10.6 — Details of the SCS mechanism for complex formation from solvento-metal ions and chelating ligands. The inset relating to k_{rc} shows competition between ring closure and solvent return.

Table 10.7 — Kinetic data relating to the SCS (sterically controlled substitution) mechanism for formation of chelate complexes; all rate constants are in units of $M^{-1} s^{-1}$, at 298.2 K in aqueous solution

Cobalt(II)			
	water exchange:	2×10^6	
	complex formation with		
	monodentate ligands:	uncharged ~1 to 3×10^5	
		charge $1-$ ~1 to 3×10^6	

5-membered rings:	glycinate⁻	α-alaninate⁻	α-aminobutyrate⁻	iminodiacetate²⁻
	2×10^6	2×10^6	2.5×10^5	1×10^7

6-membered rings:	β-alaninate⁻	β-aminobutyrate⁻	iminodipropionate²⁻
	1×10^5	2×10^4	3×10^5

Copper(II)		
	water exchange:	4×10^9
	reaction with: ammonia ⎫	2 to 20×10^8
	pyridine ⎬	
	imidazole ⎭	

5-membered-ring: α-alaninate	10×10^8
6-membered-ring: β-alaninate	2×10^8
7-membered-ring: L-carnosine[a]	5×10^4

[a]L-carnosine =

chelate ring closure has become the rate-limiting step, is sometimes called the 'sterically controlled substitution' or SCS mechanism.

10.4 POLYDENTATE AND MACROCYCLIC LIGANDS

Formation of chelate rings from reactions of metal ions with appropriate bidentate ligands may be modestly but significantly slower than formation of complexes of analogous monodentate ligands, as detailed in the preceding section. Formation reactions with analogous open-chain polydentate ligands, e.g. those depicted in Fig. 10.7(a), (b) and (c), to form products with two, three, or more chelate rings per metal ion occur at comparable rates to those with bidentate ligands, e.g. Fig. 10.7(d) or (e).

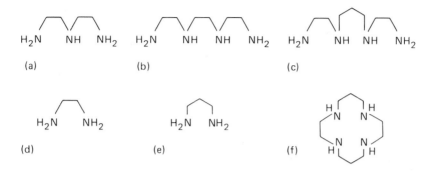

Fig. 10.7 — Polydentate garland (open-chain) and macrocyclic ligands.

If the polydentate ligand has to form large chelate rings (6-, 7- or more membered), or has steric impediments to chelate ring closure, then reaction will be rather slower. Thus, for instance, Cu^{2+} aq reacts with uncharged monodentate ligands with k_f between 10^8 and 10^9 $M^{-1} s^{-1}$ (Table 10.7), but with human serum apo-transferrin with k_f only about 10^5 $M^{-1} s^{-1}$. For this linear transferrin protein the presence of hydrophobic substituents and specific orientational requirements lead to low formation rates.

However, if the polydentate ligand is macrocyclic, then complex formation may be considerably more difficult than for open-chain analogous — compare Fig. 10.7(f) with 10.7(c). Thus for Cu^{2+} aq reacting with open-chain polydentate ligands k_f values are in the region of 10^8 $M^{-1} s^{-1}$, but rate constants are markedly less for reaction with tetraaza-macrocycles (Table 10.8), even after allowing for the $1+$ charge on the ligands in these reactions. Similar comments apply to reactions in basic solution, where the copper(II) is present in the form of $[Cu(OH)_3]^-$ (Table 10.9). The fairly strong basic properties of polyaza-ligands make it very difficult to determine rate constants specifically for formation from neutral ligands of this type plus metal ions in the M^{n+} aq form rather than as hydroxo-aqua-ions. This difficulty can be circumvented by switching to the much less basic sulphur analogues. Table 10.10 shows a large decrease in formation rate constant for tetrathia-macrocyclic ligands. It is interesting that formation rate constants here vary little with ring size, except that the ligand with the smallest ring (12 atoms) reacts markedly more slowly than the others (13 to 16 atoms). The situation is the same for tetraaza-macrocycles, where reaction with the 12-atom ring ligand is particularly slow. It seems that it is especially difficult to fold a 12-membered ring so that all four donor atoms can bond effectively to the central metal ion when this is relatively small (i.e. first-row transition metal cations). It should be said that, although the general principle that formation reactions with macrocyclic ligands are relatively slow is true enough, the detailed patterns of the

Table 10.8 — Rate constants, k_f ($M^{-1} s^{-1}$), for formation of complexes from Cu^{2+} aq and protonated cyclic tetraazamacrocycles (at 298.2 K)

Free ligand (LLLL)

R: various combinations of H and Me

k_f for Cu^{2+} aq + $LLLH^+$	4 to 40×10^6	3 to 120×10^5	8×10^3

Table 10.9 — Rate constants, k_f($M^{-1} s^{-1}$), for formation of complexes from $[Cu(OH)_3]^-$ and cyclic tetraazamacrocycles in alkaline aqueous solution (at 298.2 K)

Ligand (LLLL)			
	$R^1 = H$	Me	Me
	$R^2 = H$	H	Me
k_f for $[Cu(OH)_3]^- + LLLL$	1.0×10^{7} [a]	2.7×10^6	5.6×10^5	$\sim 10^4$

[a] Replacement of both –NH_2 groups by –NHEt leads to $k_f = 0.3 \times 10^7 M^{-1} s^{-1}$.

dependence of reactivity on ring size, on the nature of any substituents on the ligating atoms or on ring-carbon atoms, on ring flexibility, and indeed on the natures of the ligating atoms and of the metal ion, are difficult to establish and explain precisely. The advantage of having so many variables is the possibility of probing various

Table 10.10 — Rate constants, k_f ($M^{-1}s^{-1}$), for formation of macrocyclic complexes from Cu^{2+} aq and tetrathiamacrocycles; in 80% methanol at 298.2 K

Ligand (LLLL)		

k_f for Cu^{2+} aq + LLLL	$\sim 10^6$ [a]	1 to 4×10^4	0.12×10^4

[a]Estimated.

mechanistic details by appropriate choice of reagents and conditions. Thus by extensive substitution at carbon and nitrogen atoms of tetraaza-macrocyclic ligands one can make the conformational changes of the partially coordinated ligand slow enough to see as a kinetically distinct step after initial bonding of the macrocycle to, for instance, Ni^{2+}.

All the macrocyclic ligands mentioned so far have been flexible, and thus can change their conformations to accommodate the stepwise formation of metal–ligand bonds consequent on the stepwise creation of coordination site vacancies at the metal ion from consecutive loss of coordinated solvent molecules. Complex formation with a rigid multidentate ligand is much harder, as is clear from the rate constants given in Table 10.11 for a range of rigid planar tetraaza-macrocycles. Geometry, nature of substituents, and nature of the metal ion still affect reactivity, but overall k_f values are much less for these rigid ligands than for flexible macrocycles or open-chain polydentate ligands. Indeed the barriers to complex formation are really even bigger than suggested by comparing rate constants in Table 10.11 with those in earlier tables, for in most cases of complex formation from rigid macrocycles full kinetic investigations reveal that assistance is normally required from a 'synergic' anion or from a second metal ion. The latter is revealed by determination of the order with respect to the metal ion. Anion effects have been established for, for example, formation of the copper(II) complex of chlorophyllic acid a (Fig. 10.8). Rate constants vary over a hundredfold range as the nature of the anion necessarily present is varied (acetate, nitrate, etc.). Addition of ammonia, which clearly affects reactivity by complexing with the copper(II), has a similar effect on the rate constant, as does variation of solvent used.

Rate constants for complex formation with a given rigid macrocyclic ligand vary considerably with the nature of the metal ion, since cation desolvation is an important factor. The rate constants given in Table 10.12 for complex formation with tetrakis-(N-methyl)tetraphenylporphin follow the same trend as solvent exchange

Table 10.11 — Rate constants for formation of copper(II) complexes of rigid macrocyclic ligands; in aqueous solution at 298.2 K

	k_f (M^{-1} s^{-1})
deuteroporphyrin-2,4-disulpho-nic acid dimethyl ester	4.3
$R^1 = R^2 = SO_3H$	
$R^3 = R^4 = CH_2CH_2CO_2CH_3$	
haematoporphyrin IX	~0.01
$R^1 = R^2 = CH(OH)CH_3$	
$R^3 = R^4 = CH_2CH_2CO_2H$	
meso-tetraphenylporphine (5 derivatives)	0.001 to 0.02
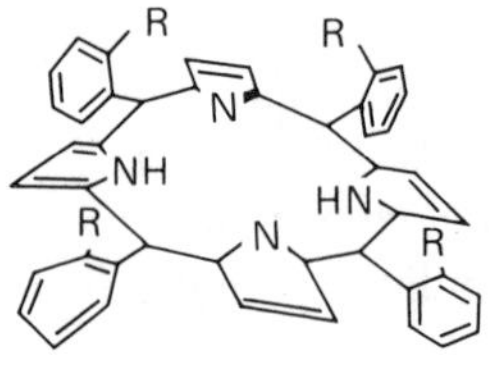 'picket fence' porphyrin 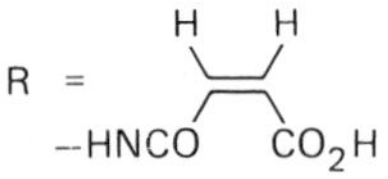	5.6[a]

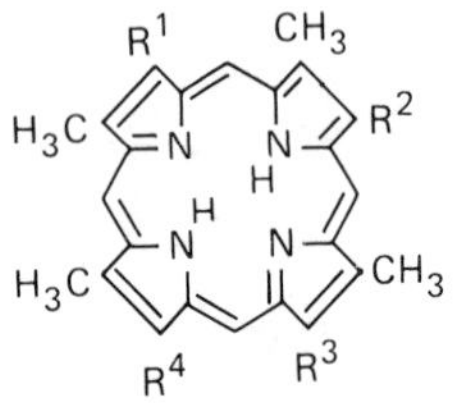

^{a}First-order rate constant (s^{-1}) for intramolecular incorporation of Cu^{2+} into the porphyrin ring, i.e. for:

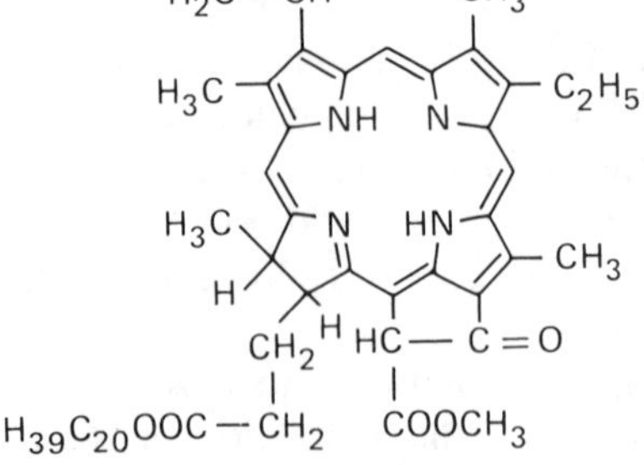

Fig. 10.8 — Chlorophyllic acid a.

Table 10.12 — Dependence of metal ion incorporation rate constants on the nature of the metal

	tetrakis(N-methyl) tetraphenyl porphine $k_f\,(\mathrm{M^{-1}\,s^{-1}})$	'picket fence' porphyrin[a] $k\,(\mathrm{s^{-1}})$
Co^{2+}	0.68	0.1
Ni^{2+}	0.0003	0.0003
Cu^{2+}	289	5.6
Zn^{2+}	10.4	1.8

[a]For the incorporation process detailed in the footnote to Table 10.11.

rate constants (Chapter 9), themselves controlled by Crystal Field Activation Energy contributions. Table 10.12 also includes rate constants for the incorporation process depicted in the footnote to Table 10.11. Reactivities for the four metal ions come in the same sequence, but the range of reactivities is much less. This is simply because the metal ions have already become partially stripped of their solvation shells on complexation by the side-chain oxygens.

10.5   CROWN ETHERS AND CRYPTANDS

The picture developed in the previous section for reactions of transition metal ions with flexible and rigid macrocyclic ligands is closely paralleled by the reactivity patterns for formation of complexes between alkali metal cations (Li^+ to Cs^+) and alkaline earth cations (Ca^{2+}, Sr^{2+}, and Ba^{2+}) and crown ethers (flexible) and cryptands (more rigid). These complex formation reactions are of biological importance through their close relation to processes of ion transport through membranes, where macrocyclic ligands of the crown ether type play a key role. The formulae and conventional names and symbols for a range of crown ethers and cryptands are shown in Fig. 10.9. This figure also includes analogous biological ligands, to show their relation to the synthetic crown ethers. The arrangement of ligating atoms is entirely analogous, but obviously other parts of the bioligands are more complicated, to fit them for their *in vivo* role.

The remarkable feature of all these ligands is that they can form particularly stable complexes with alkali metal and alkaline earth cations (Chapter 6). Crown ether complexes of the alkali metal cations are fairly unstable in water, but alkali metal cryptates can be stable even in aqueous solution; stabilities are higher in organic solvents. Stability constants are determined by ratios between rate constants for formation and rate constants for dissociation. The remarkable variations in stability constants for different metal ions reacting with a given ligand of this type are in fact determined by dissociation kinetics rather than formation kinetics (see Chapter 8). Nonetheless formation kinetics for complexes of this type show several interesting features, particularly in relation to the structure and flexibility (or otherwise) of the ligand. Interpretation of observed reactivities is by no means

Crown ethers:

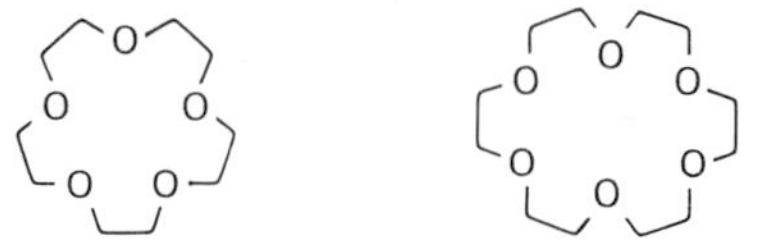

15—crown—5   18—crown—6

Cryplands:

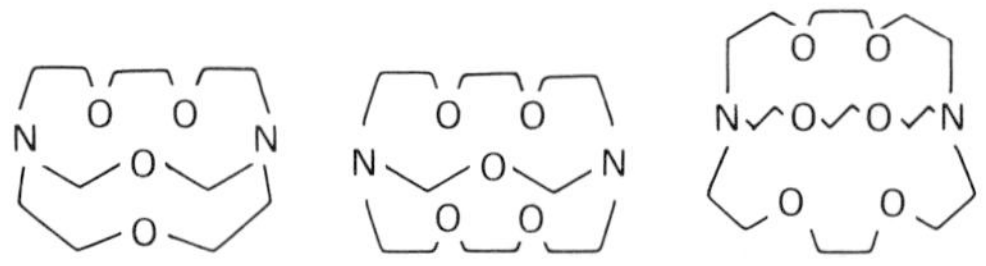

[211]   [221]   [222]

Bioligands: Valinomycin: Fig. 6:10b

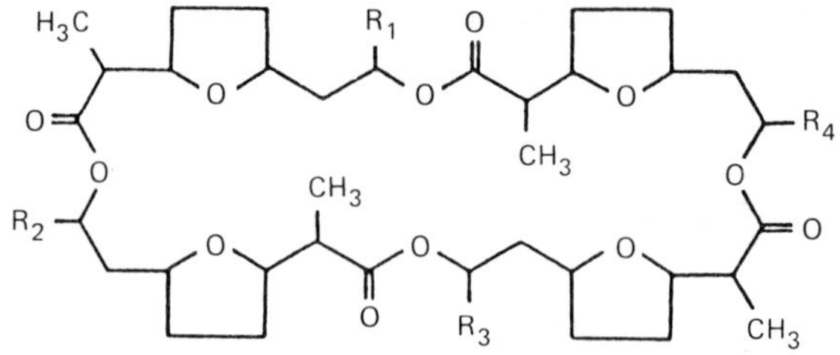

	R₁	R₂	R₃	R₄
Nonactin	Me	Me	Me	Me
Monactin	Me	Et	Me	Me
Dinactin	Me	Me	Et	Et
Trinactin	Me	Et	Et	Et

Fig. 10.9 — Synthetic and biological crown ethers and cryptands.

always easy, and several mechanistic questions still await satisfactory answers. The main difficulty is that rate constants for several steps in the overall complex formation sequence may be similar in magnitude and therefore difficult to disentangle. These steps may be two or more of the following — solvent loss from the cation, formation of outer-sphere complexes, formation of the first cation–ligand bond (i.e. the interchange step), and one or more ring closures with attendant changes in ligand conformation. There is one other process whose rate constant may be relevant, or may interfere by coupling, and that is conformational change in the free ligand.

Many years ago it was shown that rate constants for reactions of the alkali metal cations with dibenzo-30-crown-10 were much slower than expected from solvent exchange rate constants, and indeed were practically independent of the nature of the cation (Table 10.13). These observations were rationalised by suggesting that the process whose rate constant was being monitored was a conformational change in the ligand, to give an isomeric form which could react easily and quickly with the cation. Since then it has become clear that the situation is less clear-cut, with at least some

Table 10.13 — Rate constants for reaction of dibenzo-30-crown-10 with alkali metal cations; in methanol at 298.2 K.

	k_f $(M^{-1}\,s^{-1})$
K^+	6×10^8
Rb^+	8×10^8
Cs^+	8×10^8
Tl^+	8×10^8

dibenzo-30-crown-10

participation by the metal ion in each step. Indeed for such cases as complex formation between K^+ and 18-crown-6, the rate constants for the two observable kinetic processes have rather similar time constants and will be strongly coupled. For this reaction in aqueous solution it appears that the two rate constants differ by a factor of less than two. Table 10.14 lists a section of rate constants for the rate limiting

Table 10.14 — Rate constants for reaction of crown ethers with metal ions; in water at 298.2 K

	$10^{-8}k_f$ $(M^{-1}\,s^{-1})$	
	18-crown-6	15-crown-5
Li^+	~ 0.8	
Na^+	2.2	2.4
K^+	4.3	4.3
Rb^+	4.4	4.6
Ag^+	11.2	6.4
Tl^+	9.0	7.1
Sr^{2+}	0.8	0.7
Ba^{2+}	1.3	1.2
Pb^{2+}	3.3	3.2
Hg^{2+}	4.0	1.6

step in reactions between two crown ethers and a series of cations. The real but small variations in k_f values reflect a balance between, and coupling between, solvent loss and ligand conformational changes. In water it seems likely that the major contribution to the activation barrier is cation desolvation, but that in methanol or in DMF the major contribution arises from the energy required from ligand conformational changes. Where two (or more) kinetically distinct steps can be detected, they are likely to correspond to an early stage in the reaction sequence where cation desolvation and ligand conformational changes can be associated with the initial contact between the reactants, and to a later stage perhaps involving the last conformational change of the crown ether to close the last chelate ring and the associated loss of the last vestige of the primary solvation shell of the cation.

The importance of ligand conformational changes in determining reactivity have been demonstrated in experiments in which the crown ether has been made more rigid, by fusing benzene or cyclohexane rings into the crown (Table 10.15). This can

Table 10.15 — Effects of ligand rigidity on rate constants (k (s^{-1})) for the slower stage[a] in the formation of 18-crown-6 complexes of Na$^+$, in NN-dimethylformamide at 313 K

| $k =$ | 3.5×10^6 | 2×10^6 | $<1 \times 10^6$ |

[a]The fast first stage, involving initial bonding of the crown ether to the Na$^+$, has k_f between 4 and 6×10^8 M^{-1}s^{-1} for these three ligands.

be taken one step further, by studying the interaction of crown ethers with dialkylmercury compounds. Now the alkyl groups on the mercury provide further steric hindrance to ligand conformational changes, and rate constants decrease by several powers of ten.

Despite the increase in ligand connectedness and stiffness in going from crown ethers to cryptands, reactivities are remarkably similar for formation reactions of alkali metal cations with members of these two groups of ligands. This can be seen by comparing the data given in Table 10.16 here and Table 10.13 earlier. Only for the

Table 10.16 — Rate constants, k_f (M^{-1} s^{-1}), for formation of cryptates of alkali metal cations; in methanol at 298.2 K

	[211]	[221]	[222]
Li$^+$	4.8×10^5	1.8×10^7	
Na$^+$	3.1×10^6	1.7×10^8	2.7×10^8
K$^+$		3.8×10^8	4.7×10^8
Rb$^+$		4.1×10^8	7.6×10^8
Cs$^+$		$\sim 5 \times 10^8$	$\sim 9 \times 10^8$

small [211] cryptand does extra rigidity seem to result in slower complex formation. This may be compared with the very similar situation in regard to tetraaza- and tetrathia-macrocycles, set out in the previous section (see page 137). As for the formation reactions with crown ethers, so also with cryptands the observed rate constant or rate constants may represent combinations of desolvation, bond formation, and conformation. But here we can be somewhat more precise about geometry changes both for the ligand and associated with cation–ligand bond formation (Fig. 10.10). The preferred form for a free cryptand is *endo–endo*, but the cryptand needs

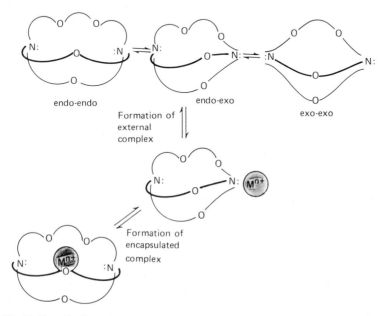

Fig. 10.10 — Conformational changes for cryptands and for cryptate formation.

to be in the *endo–exo* or *exo–exo* form before a nitrogen lone pair is available to coordinate to a cation. During complex formation the 'exclusive' form of the complex will form first, the 'inclusive' cryptate subsequently. If one designs and studies a very rigid cryptand, then the 'exclusive' → 'inclusive' change can be greatly slowed, and detected separately from the formation of the 'exclusive' form, as detailed in Table 10.17.

Reactions of alkaline earth cations with cryptands are much slower than reactions of alkali metal cations. For the alkaline earth cations, k_f values are of the order of 10^3 $M^{-1} s^{-1}$ in water at 298 K. Reaction between Cu^{2+} and cryptand [221] in dimethyl sulphoxide has a rate constant of only 37 $M^{-1} s^{-1}$. It is interesting to compare this value with data in Table 10.11. Such a comparison provides one of the very few links between this section on crown ethers and cryptands and the previous section in tetraaza-macrocycles. That section was dominated by transition metal ions, whereas this present section deals almost exclusively with *sp*-block elements. The slowness of reactions between 2 + cations and cryptands can presumably be attributed to higher energies needed for desolvation — water and dimethyl sulphoxide solvate 2 + cations

Table 10.17 — Rate constants for consecutive formation of 'exclusive' and 'inclusive' cryptate complexes of K^+

For K^+ plus

$k_1 = 3 \times 10^8 \text{ M}^{-1}\text{ s}^{-1}$

for formation of *exclusive* form

$k_2 = 8 \times 10^3 \text{ s}^{-1}$

for *exclusive* → *inclusive*

much more heavily than water and methanol solvate alkali metal cations (see section 4.3, where hydration and solvation enthalpies provide a good guide to the likely ease or otherwise of removing solvating solvent molecules in formation reactions of solvento-cations). The rate constant for formation of the [211] cryptate of Pb^{2+}, $k_f = 9.2 \times 10^5 \text{ M}^{-1}\text{ s}^{-1}$, is intermediate in magnitude, just as the solvation of the very large Pb^{2+} cation may be expected to be intermediate in extent between the smaller $2+$ cations discussed earlier in this paragraph and the $1+$ alkali metal cations.

Some rate constants for formation reactions involving alkali metal cations and biomolecules are given in Table 10.18. These rate constants vary little with the nature

Table 10.18 — Rate constants, k_f ($\text{M}^{-1}\text{ s}^{-1}$), for reactions between alkali metal cations and biologically relevant crown ethers; in methanol at 298.2 K

Valinomycin + Na^+	1.3×10^7	Na^+ + monactin	3×10^8
K^+	3.5×10^7	dinactin	5×10^7
Rb^+	5.5×10^7	trinactin	7×10^7
Cs^+	2×10^7		

of the alkali cation or of the biomolecule, and differ little from the analogous formation rate constants for crown ether complexes (Tables 10.13 and 10.14). The reaction of dinactin with Na^+ seems to involve a single rate-limiting step which is predominantly a ligand conformational change. On the other hand, ultrasonic experiments have shown two relaxations for the reaction of valinomycin with K^+. The faster process, with $k = 4 \times 10^8 \text{ M}^{-1}\text{ s}^{-1}$, is assigned to initial complexation, the slower step with $k = 1 \times 10^7 \text{ s}^{-1}$, to a later step which involves a ligand conformational change accompanied by additional ligand to cation bond formation and associated with cation desolvation. Several other cation plus biological crown ether reactions are believed to follow a similar kinetic sequence, as indeed are reactions between alkali metal cations and open-chain antibiotics such as monensin and nigericin.

In the course of this chapter we have progressed from rather straightforward formation reactions involving simple monodentate ligands to reactions involving complicated polydentate biological ligands. It should have become clear that loss of solvent molecules from the primary solvation shell of the metal cation plays a key role in the kinetics and mechanism in all cases. As the ligands become more complicated, so ring closure and conformational changes within the ligands assume increasing importance. Eventually all three factors may couple and combine, resulting either in a complicated kinetic scheme or a single kinetic step open to a variety of mechanistic interpretations. The contributing factors are clear, but the way in which they combine and interact, for complicated ligands and in systems of biological importance, has yet to be worked out in detail in many cases. However, the majority of reactions in this area can be accommodated within the general Eigen–Winkler mechanism (an extension of the Eigen–Wilkins mechanism detailed earlier in this chapter) set out in Fig. 10.11. The balance between the various rate and

Fig. 10.11 — The Eigen–Winkler mechanism for complex formation with macrocyclic and cryptand ligands.

equilibrium constants varies with the system as described in the preceding pages. Though different aspects may dominate in different reactions, the Eigen–Winkler scheme can be applied to inorganic and to biological systems, to simple macrocycles, to crown ethers, and to encapsulating ligands such as cryptates. Its only significant shortcoming is its apparent neglect of possible ligand conformational changes prior to the pre-association step of Fig. 10.11 (cf. pages 140–142 above). The importance of K^+ in bio-systems may well be connected with its role in causing conformational changes in enzymes, an extension of the principles developed in this section.

11

Kinetics and mechanisms: substitution at complex ions

11.1 GENERAL

All the reactions discussed in this chapter involve the replacement of one ligand by another — a generalisation of the special case of solvent exchange (Chapter 9) and complex formation (Chapter 10). The kinetic and mechanistic questions are the same. One wishes to know whether the mechanism is associative or dissociative, simple or complicated, and one would like to know and understand the principal factors determining reactivities. One of the most important types of substitution reactions of complexes is solvolysis (or aquation in aqueous media). Not only is this a frequently encountered reaction, but also it is important as the reverse of complex formation. The stability constant for a complex (Chapter 6) represents the balance between formation and aquation rate constants.

This chapter deals with three important types of complex ion. The first is that typified by the $[CoCl(NH_3)_5]^+$ cation. This class includes a variety of ammine- and amine-halide and related complexes, of chromium(III), rhodium(III), and iridium(III) as well as cobalt(III). The second type of complex considered here is the general class of square-planar complexes, particular of platinum(II). The third class is the rather more restricted class of complexes of the pentacyanometallate type, particularly the iron(II) series of $[Fe(CN)_5L]^{n-}$ anions. This choice of types of complex ions allows us to cover the full range of mechanistic type from fully associative (S_N2) to limiting dissociative ($S_N1(lim)$). Here, as in previous chapters, an elementary knowledge of substitution mechanisms as normally acquired through simple organic cases is assumed. Outline descriptions can be found in the Glossary; full details will be found in the texts cited in the Further Reading section at the end of this book. All of the substitutions discussed in this chapter take place slowly. They have therefore been fully studied for a long period, and are comprehensively documented. The metal ions concerned have d^3, low-spin d^6, or d^8 electron configurations. The resultant high Crystal Field stabilisation leads to high activation energies, and thus slow rates, for substitution. Not only does this mean that kinetics can be monitored by 'conventional' techniques, particularly repeat scan UV-visible spectroscopy, but also that products and reactants can be fully characterised at leisure. In recent years much information on substitution at more labile centres has been accumulated. In general the pattern of reactivities and mechanisms that has emerged for such centres parallels that described in Chapter 9 for solvent exchange.

11.2 COBALT(III) COMPLEXES

The most commonly studied reaction of cobalt(III)–ammine or amine–halide and related complexes is solvolysis, usually aquation. A typical example is

$$[CoCl(NH_3)_5]^{2+} + H_2O \rightarrow [Co(NH_3)_5(OH_2)]^{3+} + Cl^-$$

Such reactions are often carried out in slightly acidic media. In some cases the presence of acid is required to protonate the leaving ligand so that it cannot recombine, which would result in incomplete solvolysis. An example would be

$$[Co(NH_3)_5(py)]^{3+} + H_2O + H^+ \rightarrow [Co(NH_3)_5(OH_2)]^{3+} + pyH^+$$

The vast majority of substitution reactions involving replacement of one ligand by another in this type of complex actually go through an aqua-complex intermediate. So the reaction of aquation is central to the question of substitution at cobalt(III).

The inorganic kineticist is often at a great disadvantage in comparison with his organic equivalent, since the most commonly used solvent for inorganic complexes is also a good ligand. The organic chemist can usually choose an inert solvent; his inorganic counterpart often finds that his complexes only dissolve to a useful extent in water. In the common situation that water is both solvent and incoming ligand, it is not possible to obtain the full rate law for aquation, i.e. establish the order with respect to the incoming group. This is a problem encountered in Chapter 9 in respect of solvent exchange. There it was possible to use activation entropies and volumes as good guides to mechanisms. Here it is often difficult or impossible, since if the leaving group is charged, then charge separation on transition state formation will lead to a significant and confusing solvation contribution to the activation entropy and volume. In fact there have proved to be difficulties in the way of unequivocal determination of the mechanism of aquation of cobalt(III) complexes by all conceivable approaches. Only as a result of combined efforts of a large number of investigators over several decades are we now in a position to state with confidence that substitution at cobalt(III)–ammine or amine–halide and related complexes is dissociative in mechanism.

For a long time there was one apparently glaring exception to this generalisation of dissociative activation. The rate law for base hydrolysis of complexes of this type is second-order, e.g.

$$-d[CoCl(NH_3)_5^{2+}]/dt = k_2[CoCl(NH_3)_5^{2+}][OH^-]$$

The simple explanation of bimolecular nucleophilic attack of hydroxide at cobalt, i.e. an S_N2 mechanism, while fully consistent with the above rate law, seemed unattractive in the light of the otherwise universally dissociative nature of substitution at cobalt(III). The solution to the problem, first suggested in the 1930s but not fully accepted till nearly half a century later, is the conjugate base mechanism set out in the following equations:

$$[CoCl(NH_3)_5]^{2+} + OH^- \underset{\text{fast}}{\overset{K}{\rightleftharpoons}} [CoCl(NH_2)(NH_3)_4]^+ + H_2O$$

$$[CoCl(NH_2)(NH_3)_4]^+ \overset{k}{\rightarrow} \text{products}$$

The rate of formation of products is given by

$$+d[\text{products}]/dt = k[CoCl(NH_2)(NH_3)_4^+]$$

for dissociative reaction of the conjugate base complex (or indeed for associative attack by water too, since water is present in large excess!). But the concentration of the conjugate base species is given by

$$[CoCl(NH_2)(NH_3)_4^+] = K[CoCl(NH_3)_5^{2+}][OH^-]$$

so the rate law may be written

$$+d[\text{products}]/dt = Kk[CoCl(NH_3)_5^{2+}][OH^-]$$

Thus a dissociative rate-determining step can be retained despite the second-order rate law — the term in $[OH^-]$ comes from the fast pre-equilibrium. The constant K is generally extremely small, and it is not possible to detect the conjugate base intermediate, at least in aqueous media (see section 5.4).

Thus dissociative activation is the rule for cobalt(III) complexes of the ammine/amine type. When one moves away from cobalt, either across the Periodic Table to chromium(III), or down the Periodic Table to rhodium(III) and iridium(III), then substitution takes on a more associative character. This state of affairs parallels that stated for solvent exchange in Chapter 9. For chromium(III) complexes, it is easier for an incoming nucleophilic ligand to form an S_N2 transition state, as its approach is less hindered by the smaller number of d electrons around chromium(III) than around cobalt(III). The larger size of rhodium and iridium makes formation of a seven-coordinate transition state easier at these centres than at cobalt(III).

11.3 PLATINUM(II) COMPLEXES

This section is concerned with substitution at square-planar complexes. In practice such complexes are only encountered for a very small number of metal ions, almost exclusively transition metal ions of d^8 configuration. Moreover substitution is an important reaction for only some of this already small number. For low oxidation state centres such as rhodium(I) and iridium(I), the characteristic reaction is that of oxidative addition rather than substitution. For gold(III), the highly oxidising nature of the metal centre means that intended substitution reactions often turn into ligand oxidation instead. The square-planar centre of paramount importance in the study of substitution mechanisms is platinum(II). A vast number of platinum(II) complexes have been prepared and characterised, to provide a large number of potential substrates, and substitution generally takes place at slow rates which can be monitored by 'conventional' kinetic techniques. All that is said below applies equally to palladium(II) complexes, but these are much more labile and correspondingly, at least until recent years, more difficult to work with.

A typical graph of the dependence of experimentally obtained rate constants (under conditions of a large excess of incoming ligand to give first-order kinetics) on the nature and concentration of the incoming group is shown in Fig. 11.1. The rate law indicated is

$$-d[Pt^{II}\ complex]/dt = \{k_1 + k_2[nucl]\}[Pt^{II}\ complex]$$

The k_2 term corresponds to associative (S_N2) attack by the incoming ligand. Values for k_2 increase as the affinity of the incoming ligand for platinum(II) increases, just as one would expect for an S_N2 mechanism. The incoming ligand is clearly not involved in the k_1 term's transition state. Change of solvent has a large effect on k_1 values, for the solvent interacts strongly both with the leaving ligand and with the platinum (Fig. 11.2).

One further feature of platinum(II), and palladium(II) and gold(III), substitution kinetics which should be mentioned is the strong effect on reactivity exhibited by the ligand *trans* to the leaving group. It is possible to put ligands into a *trans*-effect series:

$$
\begin{array}{lll}
CO & PR_3 & SCN^- \\
CN^- & > P(OR)_3 > tu > NO_2^- > Br^- > Cl^- > py > NH_3 > OH^- > H_2O \\
CH_2{:}CH_2 & H^- & I^-
\end{array}
$$

Ligands on the left promote quick loss of a leaving group *trans* to them, while ligands on the right have little effect. The main factor determining the *trans*-effect order is π-bonding propensity, but other factors, e.g. highly polarising nature, have to be invoked in order to explain the position of ligands such as hydride in the *trans*-effect series.

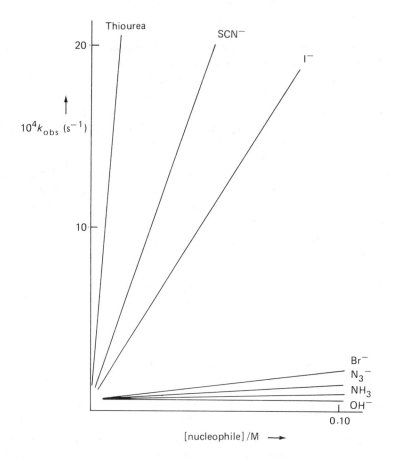

Fig. 11.1 — Kinetic pattern for substitution at $[PtCl_2(en)]$ (in aqueous solution at 308.2 K with [nucleophile] $\gg$ [Pt^{II} complex]).

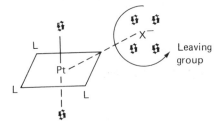

Fig. 11.2 — Solvation effects at the platinum and at the leaving group in the dissociative pathway for substitution at platinum(II) complexes (k_1 term in the rate law).

11.4 PENTACYANOFERRATES(II)

This rather specialised group of complexes is interesting in that their substitution reactions provide examples of the operation of the limiting dissociative mechanism. This mechanism, $S_N1(\text{lim})$ in organic parlance, D in inorganic, involves two stages, with a transient intermediate. The dependence of experimentally obtained rate constants (the incoming ligands in large excess) on the nature and concentration of the incoming group is shown for ligand replacement reactions

$$[\text{Fe(CN)}_5\text{L}]^{3-} + \text{L}' \rightarrow [\text{Fe(CN)}_5\text{L}']^{3-} + \text{L}$$

in Fig. 11.3(a). This pattern is characteristic of a D mechanism, as is the derived

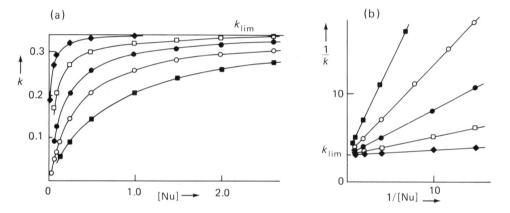

Fig. 11.3 — Kinetic pattern for the limiting dissociative, D, mechanism ([nucleophile] $\gg$ [complex]).

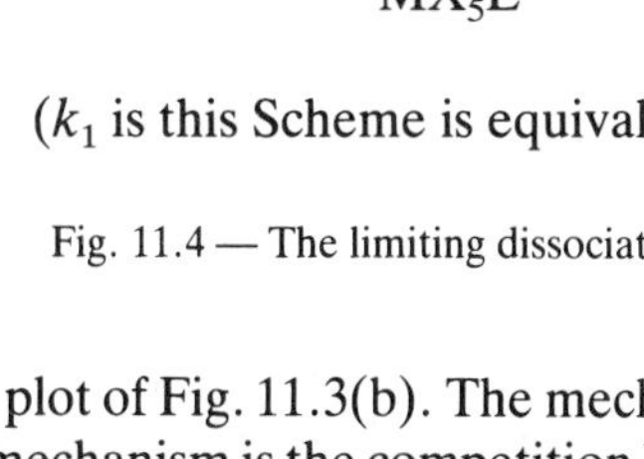

$$\text{MX}_5\text{L} \underset{k_{-1}}{\overset{k_1}{\rightleftharpoons}} \text{MX}_5 + \text{L}$$

$$\Big\downarrow k_2$$

$$+\,\text{L}'$$

$$\text{MX}_5\text{L}'$$

(k_1 is this Scheme is equivalent to k_{lim} in Fig. 11.3)

Fig. 11.4 — The limiting dissociative, D, mechanism.

double reciprocal plot of Fig. 11.3(b). The mechanistic picture is shown in Fig. 11.4. The key to the D mechanism is the competition between outgoing L and incoming L' for the transient $[\text{Fe(CN)}_5]^{3-}$ intermediate. Powerful incoming groups compete effectively with the leaving group even at low concentrations, so the plot curves sharply downwards towards the origin close to the y axis. Feeble incoming groups only capture the $[Fe(CN)_5]^{3-}$ every time it is formed when they are present in high concentration. As their concentration decreases, more and more of the times that $[Fe(CN)_5]^{3-}$ species are formed they are recaptured by the group that has just left. Thus the rate of formation of the product decreases steadily as the concentration of a weak incoming group decreases. The plots for all incoming groups will eventually reach the same limiting value, which corresponds to the rate constant for breaking the iron to leaving L bond in the starting $[Fe(CN)_5L]^{3-}$ anion.

The D mechanism is relatively rare for inorganic substitutions. It is not often that loss of a ligand from a complex leaves a species stable enough to persist long enough to discriminate between leaving and potential entering groups. In the case of the pentacyanoferrates(II), the very high ligand field of the cyanides seems to be just enough for five of them to confer limited life on $[Fe(CN)_5]^{3-}$. It is interesting to note that the next metal, cobalt, only forms a pentacyano-complex in the 2+ oxidation state — $[Co(CN)_5]^{3-}$ and not $[Co(CN)_6]^{4-}$.

The D mechanism does operate in a number of important biochemical systems. Such entities as haem have iron with four very strongly bound nitrogen atoms from a macrocyclic ring. If the fifth coordination position has another firmly bonded ligand, such as histidine–nitrogen from another part of the enveloping protein, then ligand replacement reactions may take place at the sixth coordination site by the D mechanism. Several instances are known at this type of iron site, and others at other metals such as cobalt.

11.5   OTHER COMPLEXES

For discussions of kinetics and mechanisms of substitution at the numerous metal centres not mentioned in the preceding three sections, and indeed for much more detail concerning the kinetics and mechanisms of substitution at cobalt(III), platinum(II), and iron(II), the reader is referred to the various texts cited in the Further Reading section at the end of this book.

12

Kinetics and mechanisms: redox reactions

12.1 INTRODUCTION

The final major topic under consideration in this book is that of the kinetics and mechanisms of electron transfer, or redox, reactions involving ions in solution. Fig. 12.1 includes a varied selection of redox reactions involving pairs of ions, arranged according to type of reagent and, within each type, ordered according to rate constant (in aqueous solution). The examples given cover a wide range of simple and of complex ions, and a wide range of reactivities. Reactions of aqua-metal ions with hydrated electrons are usually very fast, with rate constants approaching the diffusion controlled limit. A large number of inorganic reactions are, though much less rapid than this, still fast by conventional standards. Under normal concentration conditions at least half of the reactions cited in Fig. 12.1 need to be followed by fast reaction techniques such as stopped-flow or T-jump methods. On the other hand, some oxidations, for example by perchlorate, peroxodisulphate, or some cobalt(III) complexes, proceed remarkably slowly. In this chapter we shall be considering both reactivities and mechanisms, and the factors controlling them. The mechanistic aspect will be discussed with particular reference to redox reactions between pairs of transition metal complexes, including the important special case of aqua-metal cations, since the most convincing and detailed evidence is available in this area.

In essence there are two mechanisms for electron transfer, the so called 'inner-sphere' and 'outer-sphere' mechanisms. These are outlined in Fig. 12.2, and described in detail in the following two sections. The outer-sphere mechanism involves simply the transfer of an electron, with no change in the primary coordination shell of the central atom of either reactant, be it a metal complex or a polyatomic anion. The inner-sphere mechanism involves substitution as an essential prerequisite for electron transfer, and often involves atom or group transfer as well. The simplicity of the outer-sphere mechanism makes it difficult to obtain convincing *proof* of its operation in a given system, whereas the essential role of substitution in the inner-sphere mechanisms mean that it is possible, in favourable cases, to obtain very convincing

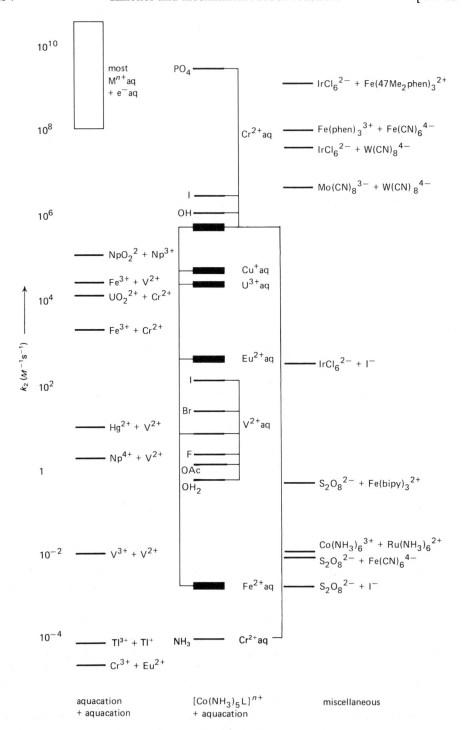

Fig. 12.1 — Second-order rate constants for selected redox reactions involving metal complexes and aqua-metal cations. All data refer to aqueous solution, 298.2 K; the thick bars in the middle column all refer to reduction of $[CoCl(NH_3)_5]^{2+}$ by various aquacations.

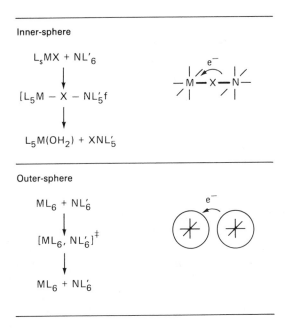

Inner-sphere

$$L_sMX + NL'_6$$

$$\downarrow$$

$$[L_5M - X - NL'_5f$$

$$\downarrow$$

$$L_5M(OH_2) + XNL'_5$$

Outer-sphere

$$ML_6 + NL'_6$$

$$\downarrow$$

$$[ML_6, NL'_6]^{\ddagger}$$

$$\downarrow$$

$$ML_6 + NL'_6$$

Fig. 12.2 — The inner-sphere and outer-sphere electron transfer mechanisms. The left-hand side shows the constituents of the primary coordination shell of each metal ion (charges are omitted in the interests of clarity); the right-hand side indicates the modes of electron transfer in the respective transition states.

evidence for its operation. We shall therefore deal with the inner-sphere mechanism first. Discussion of the outer-sphere mechanism then follows naturally, in the sense that the best established outer-sphere reactions are those where the inner-sphere pathway can most clearly be ruled out.

12.2   THE INNER-SPHERE MECHANISM

This mechanism was described by Taube in 1954. The choice of system was based on his experience of substitution inertness and lability of transition metal complexes, built up in the years preceding in the process of compiling the first comprehensive review of kinetics of substitution at transition metal centres. As will become apparent, the choice of metal centres is critical to the gaining of conclusive proof of the concurrent substitution and electron transfer components of this mechanism. Taube's key experiments involved hexaaquachromium(II) reduction of the classic Werner complex $[CoCl(NH_3)_5]^{2+}$. The main features are set out in Fig. 12.3. The characterisation of the chromium(III) product as $CrCl^{2+}aq$ rather than $Cr^{3+}aq$ immediately suggests that something more than simple electron transfer has taken place. Detailed consideration of the kinetic characteristics of the reactants and products proves, insofar as any mechanism can be proved, that transfer of chloride

Reaction: $CoCl(NH_3)_5{}^{2+} + Cr^{2+}aq \xrightarrow{\text{acid}} CrCl^{2+}aq + Co^{2+} + 5NH_4{}^+$				
Substitution time-scale	hours	10^{-9} s	hours	10^{-6} s
Redox time-scale		fraction of a second		

Fig. 12.3 — Taube's original inner-sphere redox reaction, showing the approximate time scales involved.

from cobalt to chromium is an essential feature of the redox process. The kinetic details are included in Fig. 12.3; the important considerations here are the relations between the half-lives characteristic of substitution at the reactant and product complexes and the time-scale for the redox reaction, which is of the order of milliseconds. Clearly the cobalt(III) must enter the redox transition state with the chloride still bonded to the cobalt. But likewise the chloride must become bonded to the chromium before the transition state separates into products. The half-life for the $Cr^{3+}aq$ plus Cl^- reaction is so long that there would be negligible incorporation of chloride into the coordination shell of the chromium(III) within the time needed for product characterisation if the primary oxidation product had been $Cr^{3+}aq$. The only possible conclusion is that *both* the cobalt and the chromium must have been bonded to the chloride at the instant of electron transfer, i.e. in the transition state (the Frank–Condon principle states that electron transfers take place on a much shorter timescale than bond making or breaking). It should be added that substitution at $Cr^{2+}aq$ takes place extremely rapidly, so there is no significant barrier to the displacement of one of its water molecules by the incoming bridging chloride. In similar vein, cobalt(II) is substitution labile, so it detaches itself rapidly from the bridging chloride as soon as electron transfer has taken place. Moreover the $[Co(NH_3)_5]^{3+}$ moiety thereby produced also rapidly goes to $Co^{2+}aq$ and ammonia, the latter scavenged instantly by the acid present to give NH_4^+ as ultimate product. This inner-sphere mechanism can be summarised pictorially as shown in Fig. 12.4.

The convincing demonstration just discussed depends on the fortunate coincidence of involving two inert complexes — one a reactant, the other a product. If the products are all labile, then obviously definitive proof of ligand transfer is impossible. Recent improvements in techniques have enabled the transfer of ligands from cobalt(III) to iron(II) to be demonstrated, since iron(III) is only moderately labile. But europium(III), vanadium(III), and many other such oxidised forms are too labile for ligand transfer to be proved, though in some cases it can be inferred indirectly.

A wide variety of ligands can act as bridging ligands in electron transfer reactions between metal centres. The main requirement is that the potentially bridging ligand has, when coordinated to one metal centre, a lone pair of electrons free and available to bond to the other metal centre. This is generally the case for the Group VII ligands F^-, Cl^-, Br^-, and I^-, and for ligands containing Group VI donor atoms such as oxygen or sulphur. But Group V donors, such as ammonia, pyridine, amines, or

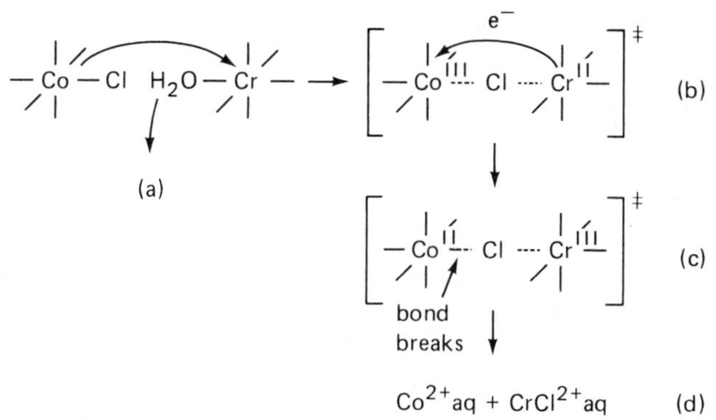

Fig. 12.4 — The inner-sphere electron transfer mechanism, for chromium(II) reduction of a chloro-cobalt(III) complex.

phosphines, do not have a lone pair free on the donor atom for bridge formation. Such ligands can only act as bridges if they have a lone pair available elsewhere in the molecule or ion — pyrazine (N⟮ ⟯N) or azide (N_3^-) can act as bridging ligands for inner-sphere electron transfer, whereas pyridine or ammonia cannot. The great advantage of the inner-sphere route is demonstrated by comparing, for instance, rate constants for reduction of $[CoCl(NH_3)_5]^{2+}$ and of $[Co(NH_3)_6]^{3+}$ by chromium(II). This and related comparisons are set out in Table 12.1. For each reducing aqua-metal

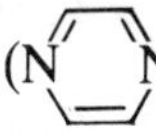

Table 12.1 — Rate constants (k_2 ($M^{-1} s^{-1}$)) for aqua-metal cation reductions of cobalt(III) complexes, in aqueous solution at 298.2 K

	Cr^{2+}	V^{2+}	Eu^{2+}	Cu^+	U^{3+}
Inner-sphere possible:					
$[CoCl(NH_3)_5]^{2+}$	6×10^5	7.6	400	4.9×10^4	3.2×10^4
$[CoBr(NH_3)_5]^{2+}$	1.4×10^6	25	250	4.5×10^5	1.4×10^4
$[CoI(NH_3)_5]^{2+}$	3×10^6	120	120		
$[Co(OH)(NH_3)_5]^{2+}$	1.5×10^6			3.8×10^2	
Inner-sphere impossible:					
$[Co(NH_3)_6]^{3+}$	9×10^{-5}	3.7×10^{-3}	1.7×10^{-3}		1.3
$[Co(en)_3]^{3+}$	2×10^{-5}	2×10^{-4}	8×10^{-4}	$\leqslant 4 \times 10^{-4}$	0.13

ion, reaction is several orders of magnitude slower when there is no potential bridging ligand present in the complex oxidant and only direct outer-sphere electron transfer is possible.

Although inner-sphere electron transfer usually results in transfer of the bridging ligand, this is not a necessary requirement. If we consider hexachloroiridate oxidation of pentacyanocobaltate(II), then the binuclear species subsequent to electron transfer is $[(NC)_5Co^{III}-Cl-Ir^{III}Cl_5]^{5-}$. Here both metals are low-spin d^6 and thus inert to substitution and reluctant to dissociate from the bridging chloride. The third-row element iridium is attached even more firmly to this chloride than the first-row cobalt, so the post-electron transfer binuclear species comes apart at the Co–Cl bond. The fact that the bridging chloride then remains bonded to the iridium means that no ligand transfer has taken place. Another example of inner-sphere electron transfer not accompanied by ligand transfer is provided by the chromium(II) reduction of $[Ru(NH_3)_5(isonicotinamide)]^{3+}$. After the ruthenium has gained its electron it is in the low-spin d^6 state. This has an even higher Crystal Field stabilisation than the d^3 chromium(III), so the isonicotinamide stays with the ruthenium after electron transfer. Inner-sphere electron transfer without bridging ligand transfer is rare; even such closely related reactions as chromium(II) reduction of $[IrCl_6]^{2-}$ and $[Co(CN)_5]^{3-}$ reduction of $[IrBr_6]^{2-}$ go by parallel ligand transfer and no-ligand transfer paths (indeed the $[IrCl_6]^{2-}$ plus Cr^{2+} aq reaction also involves a third parallel path, that of outer-sphere electron transfer).

The complicated $[IrCl_6]^{2-}$ plus Cr^{2+} aq redox reaction has yet another feature of interest. On mixing the two solutions, a transient deep blue colour may be observed. In practice the inner-sphere transition state is not that, but is an intermediate of significant lifetime. Only slightly stronger bonding between two metal centres and the bridging ligand in a post-electron transfer state means that such a binuclear species can be isolated and characterised. This is the case in the following examples:

$$[Co(CN)_5]^{3-} + [Fe(CN)_6]^{3-} \rightarrow [(NC)_5Co^{III}-NC-Fe^{II}(CN)_5]^{6-}$$
$$UO_2^{2+} aq + Cr^{2+} aq \rightarrow [O=U=O-Cr]^{4+} aq.$$

In the first case, where the binuclear anion can be isolated in the form of its potassium or barium salt, both the cobalt(III) and the iron(II) are low-spin d^6, and thus substitution inert through crystal field stabilisation. In the latter case it is the 5+ formal charge on the uranium which makes U–O bond breaking as difficult as $Cr^{III}(d^3)$–O bond breaking.

To return to simple one-step inner-sphere electron transfer, the next feature to consider is that several potential bridging ligands offer more than one site for attachment to a reductant such as Cr^{2+} aq. The alternative sites of attack are indicated for cyanide, thiocyanate, nitrite, and oxalate in Fig. 12.5. In the case of

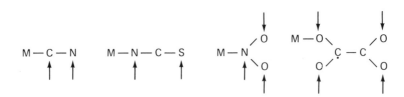

Fig. 12.5 — Likely sites of attack of a metal ion reductant (or oxidant) at coordinated ligands.

cyanide, Cr^{2+} aq could approach the nitrogen of the potentially bridging cyanide ligand in $[Co(CN)(NH_3)_5]^{2+}$ more easily than the carbon, but a shorter path for electron transfer would be available via the latter mode of attachment. In practice it has been found that this redox reaction takes place in three kinetically distinct steps, which take seconds, minutes, and many hours respectively. The known rate constants for isomerisation of $[Cr(NC)(OH_2)_5]^{2+}$ to the more stable linkage isomer $[Cr(CN)(OH_2)_5]^{2+}$, and for aquation of the latter, mean that it is possible to assign steps two and three to these reactions. Thus the first step in the sequence must represent attack of the Cr^{2+} aq at the nitrogen of the cyanide coordinated to the cobalt in the starting complex:

$$[Co(CN)(NH_3)_5]^{2+} + Cr^{2+} aq \xrightarrow{acid} [Cr(NC)(OH_2)_5]^{2+} + Co^{2+} aq + 5NH_4^+$$

$$[Cr(NC)(OH_2)_5]^{2+} \rightarrow [Cr(CN)(OH_2)_5]^{2+}$$

$$[Cr(CN)(OH_2)_5]^{2+} \xrightarrow{acid} Cr^{3+} aq + HCN$$

Similar evidence and reasoning suggest that remote attack also predominates in other cases; ease of approach of the reductant is dominant over length of electron transfer path (which is usually delocalised anyway). However, there are a few instances where adjacent attack is favoured over remote attack, generally by the tendency of a 'soft' reductant to seek out a 'soft' centre in a potential bridging ligand. Thus in reduction of $[Co(SCN)(NH_3)_5]^{2+}$ by $[Co(CN)_5]^{3-}$, the reductant bonds almost exclusively to the adjacent sulphur rather than the remote and accessible nitrogen of the coordinated thiocyanate.

12.3   THE OUTER-SPHERE MECHANISM

In contrast to the inner-sphere mechanism, it is well-nigh impossible to get totally unequivocal evidence for the operation of the outer-sphere mechanism. Where electron transfer between two complexes is much faster than substitution rates at either, then the outer-sphere mechanism is strongly indicated. Several of the very fast redox reactions included in Fig. 12.1 come into this category. Thus comparisons of substitution and redox timescales given in Table 12.2 indicates that $[Fe(CN)_6]^{4-}$

Table 12.2 — Substitution inertness of some outer-sphere redox partners (water; 298.2 K)

		k (s^{-1})	Approx. $\tau_{1/2}$
$[Fe(CN)_6]^{4-}$	for cyanide exchange	$< 10^{-6}$	> 1 week
$[Fe(CN)_6]^{3-}$	for cyanide exchange	$< 10^{-6}$	> 1 week
$[Fe(phen)_3]^{2+}$	for phen loss	7×10^{-5}	3 hours
$[Fe(4,7\text{-}Me_2phen)_3]^{2+}$	for 4,7-Me$_2$phen loss	2×10^{-5}	10 hours
$[IrCl_6]^{3-}$	for aquation	9×10^{-6}	1 day

reduction of $[Fe(phen)_3]^{3+}$ and the almost diffusion controlled $[IrCl_6]^{2-}$ oxidation of $[Fe(4,7-Me_2phen)_3]^{2+}$ cannot by any reasonable stretch of the imagination proceed by inner-sphere pathways. Other commonly invoked examples of redox reactions where marked substitution inertness makes outer-sphere electron transfer well-nigh certain are the electron exchange reactions between MnO_4^- and MnO_4^{2-} and between ferrocene and the ferrocinium cation. While it is conceivable that oxide bridging might be possible in the former — five-coordinate manganese(VI) or (VII) does not seem totally unreasonable — it is very difficult to see how a ligand bridge could be formed in the latter case (Fig. 12.6).

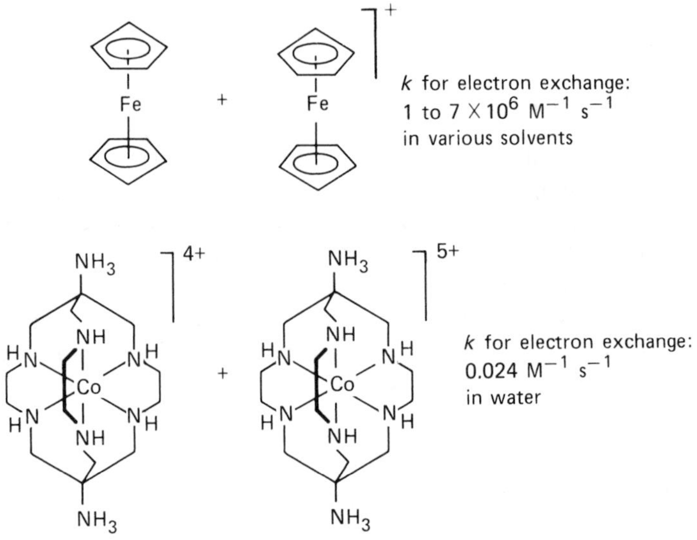

Fig. 12.6 — Two systems in which fast electron transfer must surely take place by the outer-sphere mechanism, ferrocene/ferrocinium and cobalt(II)/(III) encapsulated complexes.

The outer-sphere redox reactions discussed in the preceding paragraph are all fast. Delocalised ligands such as cyanide, 1,10-phenanthroline and its derivatives, and cyclopentadienyl permit easy movement of an electron between the central metal and the periphery of the complex. If the ligands are saturated and do not offer a delocalised π-orbital route across the ligand, then the resulting barrier to electron transfer leads to much lower rate constants. Even with delocalised ligands, electron transfer to cobalt(III) can be difficult, and therefore slow, since if the electron approaches the metal through a π-system there is a symmetry barrier to its transfer to the d_{z^2} level, its final destination. For these and other reasons, outer-sphere electron transfer can take place at very slow rates as well as the very fast rates mentioned earlier.

The outer-sphere mechanism has proved attractive to the theoretician. Its simplicity, especially the very limited extent of orbital overlap, means that it can be treated in a quantitative manner with far fewer assumptions and approximations than usual. In particular, it is reasonable to expect some correlation between kinetic and thermodynamic parameters in such systems (see Chapter 8). Indeed for many series of related reactions there is a correlation between $\Delta G^{\ddagger}$ and $\Delta G^{\ominus}$. This is illustrated, in a way more directly related to experimental observations, in the form of a log k versus redox potential plot in Fig. 12.7. This connection has been

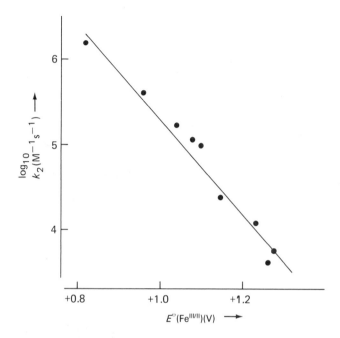

Fig. 12.7 — Correlation between logarithms of rate for cerium(IV) oxidation of substituted $[Fe(Xphen)_3]^{2+}$ cations (X = Me_4, Me_2, Me, H, Ph, Cl, SO_3^-, NO_2) in aqueous sulphuric acid and redox potentials for the iron(III)/(II) couples of the respective complexes.

developed and generalised, by Marcus, Hush, Sutin, and many others. The resultant Marcus–Hush theory of electron transfer processes relates kinetics to thermodynamics via rate constants for self-exchange reactions as well as for the electron transfer cross-reaction and the equilibrium constant for the cross-reaction. The equilibrium constant is related to the redox potentials for the two couples concerned through the Gibbs free energy for the electron transfer reaction. These quantities are linked by the normal thermodynamics equations. The theory in its barest outline is shown in Fig. 12.8.

To reverse the approach of the previous paragraph, the conformance or otherwise of a series of redox reactions to the Marcus–Hush theory is sometimes used as a test for the operation of the outer-sphere mechanism. As suggested at several points earlier in this chapter, it is only possible to make a confident assignment of electron transfer mechanism in favourable cases. There are many electron transfer reactions between metal complexes where assignment of mechanism is not straightforward. In such cases a range of circumstantial evidence has to be pressed into service in

OUTER-SPHERE ELECTRON TRANSFER

MARCUS CROSS-CORRELATION

The rate constant k_{12} for the outer-sphere redox reaction

$$ML_6^{m+} + ML_6'^{(m-1)+} \rightleftharpoons NL_6^{(m-1)+} + NL_6'^{(n+1)+}$$

can be estimated from the 'self-exchange' rate constants (see adjacent panel) for

$$ML_6^{m+} + ML_6^{(m-1)+} \quad : \text{ rate constant } k_{11}$$
$$NL_6'^{n+} + NL_6'^{(n+1)+} \quad : \text{ rate constant } k_{22}$$

and the Gibbs free energy ($\Delta G_{12}^{\ominus}$) or equilibrium constant (K_{12}) for the overall reaction from:

$$k_{12} = (k_{11} k_{22} K_{12} f)^{1/2} w \ .$$

In this equation:
(i) f is a factor which under certain conditions can be calculated from

$$\log f = \frac{(\log K_{12})^2}{4 \log(k_{11} k_{22} / z^2)}$$

where z is the encounter rate constant (ca. $10^{11} \, M^{-1} s^{-1}$). For $\Delta G_{12}^{\ominus}$ not too far from zero, f is approximately unity.
(ii) w is a work term representing the algebraic sum of the energies expended or liberated in bringing the reactants together in the self-exchange and cross reactions. w is related to the respective ion-pairing or association constants.

K_{12} ($\equiv \Delta G_{12}^{\ominus}$) can sometimes be measured directly, but is more usually obtainable from redox potential data ($\Delta G^{\ominus} = -n \mathscr{F} E^{\ominus}$; Chapter 7). The rate constant k_{12} is generally measured directly by conventional or fast reaction (stopped flow; T-jump) techniques.

In the illustration shown in Fig. 12.7, ML_6^{m+} and $ML_6^{(m-1)+}$ are cerium(IV) and cerium(III), $NL_6'^{n+}$ and $NL_6'^{(n+1)+}$ are the iron(II) and iron(III) forms of a series of ligand-substituted tris-1,10-phenanthroline complexes.

ELECTRON 'SELF-EXCHANGE'

$$ML_6^{m+} + {}^*ML_6^{(m-1)+} \rightleftharpoons ML_6^{(m-1)+} + {}^*ML_6^{m+}$$

Self-exchange rate constants k_{11} (k_{22}) can sometimes be measured directly, e.g. from n.m.r. line-broadening or isotopic exchange experiments. Alternatively they can be estimated via

$$k_{11} = (RT/h) \exp(-\Delta G_{11}^{*}/RT)$$

since the component parts of the overall Gibbs free energy of activation can themselves be estimated:

$$\Delta G_{11}^{*} = w_{11} + \Delta G_{its}^{\ddagger} + \Delta G_{in}^{\ddagger} + \Delta G_{out}^{\ddagger}$$

Here w_{11} is the work term for bringing the reactants together, calculable by simple electrostatics from their charges and dimensions:
$\Delta G_{its}^{\ddagger}$ is the intrinsic Gibbs free energy of activation;
$\Delta G_{in}^{\ddagger}$ arises from changes in bond lengths on going from the initial to the transition state; and
$\Delta G_{out}^{\ddagger}$ arises from solvation changes on going from the initial to the transition state:

USEFULNESS OF THE MARCUS CROSS-CORRELATION

(i) Estimation of k_{11} (k_{22}) values where these are impossible to measure directly.
(ii) Derivation of redox potentials which cannot be measured directly: $E^{\ominus}$ from $G_{12}^{\ominus}$ from K_{12} from measured rate constants k_{12}; k_{11}; k_{22}.

Fig. 12.8 — The Marcus–Hush theory of outer-sphere electron transfer.

attempting to establish the likely mechanism. In such circumstances the arguments used are often complex and always indirect; they may be found in specialised texts devoted to inorganic reaction mechanisms.

12.4 INTERMEDIATES, PRE-EQUILIBRIA, AND OTHER COMPLICATIONS

Readers wanting only the simplest picture of the inner- and outer-sphere mechanisms for electron transfer can restrict themselves to sections 12.1 to 12.3 above. The detailed picture of these mechanisms is rather more complicated, as will emerge in the course of this section. Anyone interested in probing more deeply into the factors determining redox reactivities, in electron transfer in real or model bioinorganic systems, or in medium effects, should be at least acquainted with the fuller picture developed in the following paragraphs.

Full details of the inner-sphere and outer-sphere pathways are set out side by side in Fig. 12.9. The initial outer-sphere association equilibrium brings the reactants into

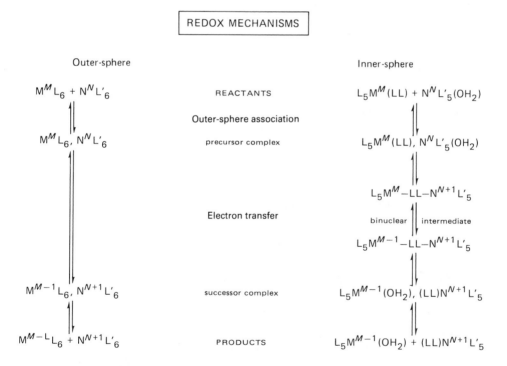

Fig. 12.9 — Details of the individual steps comprising the inner-sphere and outer-sphere electron transfer mechanisms.

close proximity to facilitate electron transfer. This pre-equilibrium, characterised by its equilibrium constant K_{os}, can be important in determining reactivity. In general the more negative the value of the charge product of the reactants, MN– with the symbols used in Fig. 12.9, the higher the value of K_{os} and therefore the easier and faster the overall redox reaction. When outer-sphere reactions are discussed in terms of the Marcus–Hush theory (see the previous section), then this pre-equilibrium is included as a work term, w_{ij}. For reactions between ions, w_{ij} may be of a similar order of magnitude to the other terms in the Marcus equation. The importance of w_{ij}, as important to inner-sphere as to outer-sphere processes, can be gauged through its reflection in measurement activation entropies and activation volumes. In the absence of solvation effects, for a bimolecular redox reaction both $\Delta S^{\ddagger}$ and $\Delta V^{\ddagger}$ should be negative; $\Delta V^{\ddagger}$ would be expected to be about $-10 \, cm^3 \, mol^{-1}$. For oxidation of $[Co(terpy)_2]^{2+}$ by $[Co(bipy)_3]^{3+}$, $\Delta V^{\ddagger} = -9.4 \, cm^3 \, mol^{-1}$ ($\Delta S^{\ddagger}$ is large and negative). These cations are large and have hydrophobic peripheries, so solvation effects are likely to be small. Reactions between pairs of positively or negatively charged ions also often have activation volumes in the region of $-10 \, cm^3 \, mol^{-1}$ (Table 12.3), but redox reactions between pairs of oppositely charged

Table 12.3 — Activation volumes for some redox reactions between ions in aqueous solution

	$\Delta V^{\ddagger}(cm^3 \, mol^{-1})$
Like-charged reactants:	
$[Co(terpy)_2]^{2+} + [Co(bipy)_3]^{3+}$	-9
$[Mo_2O_4(OH)(edtaH)]^{2-} + [IrCl_6]^{2-}$	-12
$[Fe(OH_2)_6]^{2+/3+}$ electron exchange	-12
$[Co(en)_3]^{2+/3+}$ electron exchange	-20
$[MnO_4]^{2-/-}$ electron exchange	-21
Unlike-charged reactants:[a]	
$[CoCl(NH_3)_5]^{2+} + [Fe(CN)_6]^{4-}$	$+28$
$[Co(dmso)(NH_3)_5]^{3+} + [Fe(CN)_6]^{4-}$	$+23$

[a] See also Table 12.4.

complex ions generally have large positive activation volumes (Table 12.3 again). This can be understood in terms of solvation changes. Thus if the reactants have charges M+ and M− , then the precursor complex will have zero charge. There will be considerable desolvation of the charged reactants as they form the uncharged transition state. This release of electrostricted solvent from the solvation shells of the reactants will result in an increase in both the volume and the entropy of the system.

For some reactions between complexes of opposite charge type it is possible to split the observed activation volume into outer-sphere association (ion-pairing) and electron transfer components. Table 12.4 shows examples of this, with the expected positive contribution from the preliminary ion-pairing. But it also shows that the electron transfer step tends to show a positive $\Delta V^{\ddagger}$, which is generally in the region of $+20$ to $+30 \, cm^3 \, mol^{-1}$. This too may be a solvation change effect. Volume profiles

Table 12.4 — Ion association and electron transfer components of observed (overall) volumes of activation for redox reactions between oppositely charged complexes

Reaction	$\Delta V^{\ddagger}_{obs}$	$= \Delta V^{\ominus}_{os}$	$+ \Delta V^{\ddagger}_{et}$
$[Co(NH_3)_5(py)]^{3+} + [Fe(CN)_6]^{4-}$	$+ 47$	$+ 23$	$+ 24$
$[(H_3N)_5CoO_2Co(NH_3)_5]^{5+} + [Mo_2O_4(edta)]^{4-}$	$+ 36$	$+ 24$	$+ 12$

suggest that such transition states are roughly halfway between reactants and products, so a knowledge of the partial molar volumes of the products is required for interpretation of $\Delta V^{\ddagger}$. There are a number of electron exchange or transfer reactions whose activation volumes cannot be fitted neatly into this pattern. Most of these involve highly charged ions, e.g. $[Fe(CN)_6]^{4-}$. Ion-pairing between these and redox-inactive ions present may play an important part. The effects of Li^+, Na^+, and K^+ on $[Fe(CN)_6]^{3-/4-}$ electron exchange rate constants and activation volumes shows just how large an effect ion-pairing with simple ions may have, and counsels caution in the interpretation of activation volumes determined at the moderate ionic strengths generally employed.

After the common precursor complex has been formed, the inner-sphere and outer-sphere paths diverge. In the outer-sphere mechanism there is simply electron transfer, but the inner-sphere mechanism involves several further individual steps, any of which may have an important effect on reactivity. The next step in the inner-sphere pathway is effectively a complex formation reaction, with the 'incoming ligand' L_5MLL replacing the water molecule in $NL'_5(OH_2)$. This is a special case of the general Eigen–Wilkins mechanism discussed in Chapter 10. We have already discussed the question of the lifetime of the binuclear transition state or intermediate thus formed (see middle of section 12.2, p. 158). Lastly there is a further possibility, not included in Fig. 12.9, of the electron dwelling for a time on the bridging ligand rather than moving 'instantaneously' from metal centre N to metal centre M. This situation is uncommon, since there are two restricting requirements. The first is that there is a large barrier to putting the electron into a vacant orbital on the oxidant metal ion. The practice this means a restriction to cobalt(III), where the transferred electron has to go into an e_g orbital ($t^6_{2g} \to t^6_{2g}e^1_g$). As the electron normally arrives via a π-symmetry orbital of the bridging ligand, there is a large symmetry hurdle to overcome. This hurdle does not exist, for instance, ruthenium(III), where the transferred electron can go into a t_{2g} orbital ($t^5_{2g} \to t^6_{2g}$). The second requirement is for the bridging ligand to be reducible, otherwise there is no incentive for the electron to transfer to it from the reductant metal ion. Examples of such bridging ligands include several pyrazine derivatives; such radical ligand bridged binuclear species are shown in Fig. 12.10.

12.5 METAL ION OXIDATION OF SIMPLE SPECIES

The question of the nature and lifetime of intermediates, just discussed in relation to redox reactions between metal complexes, is also important in reactions involving

Bridging radical ligand

isonicotinamide

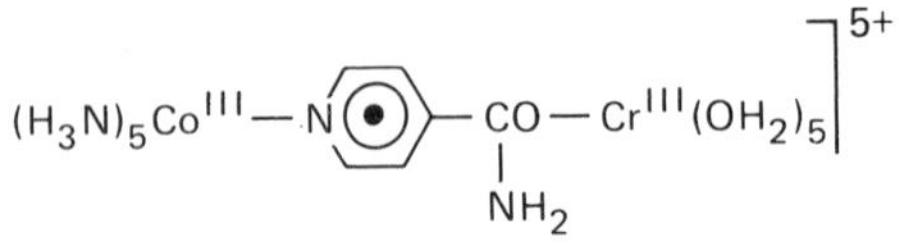

pyrazine carboxylate

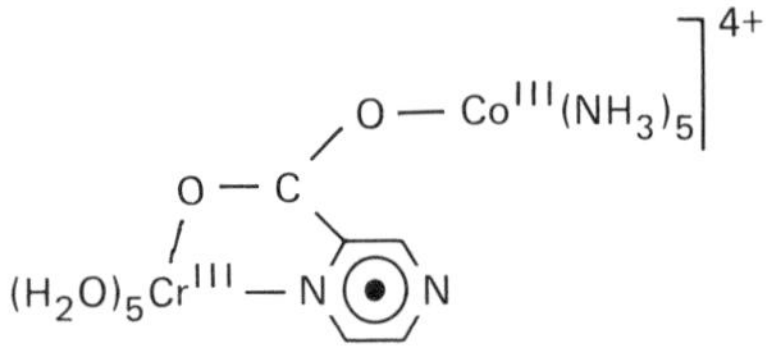

phenyl glyoxalate

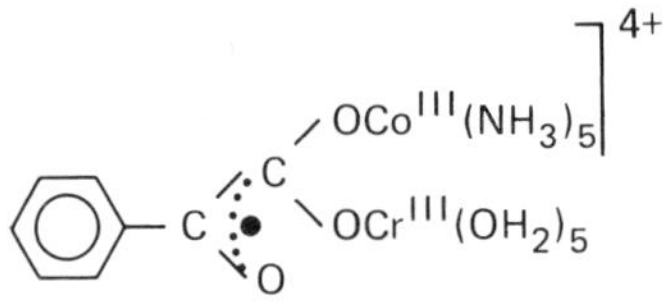

Fig. 12.10 — Examples of radical ligand bridged intermediates in inner-sphere electron transfer.

the oxidation or reduction of simple inorganic or organic ions or molecules by metal ions. Metal ion reductions, for instance by Cr^{2+}aq, V^{2+}aq, or Ti^{3+}aq, have been little studied, but there is a wealth of information on oxidations by such cations as Ce^{4+}aq, Mn^{3+}aq, Fe^{3+}aq, and Tl^{3+}aq. The main interest lies in whether the redox reaction is a simple one-step process or whether it involves complex formation with subsequent electron transfer (and sometimes atom transfer too) betwen ligand and metal:

$$M^{n+}aq + L \rightleftharpoons ML^{n+}aq$$

$$ML^{n+}aq \rightarrow \text{redox products}$$

The observed reaction pattern and kinetics are determined by the relations between the three rate constants involved.

In a few very favourable cases, for example thallium(III) oxidation of certain alkenes, it has proved possible to isolate and characterise an intermediate complex. More commonly the intermediacy of a metal ion–substrate complex is inferred from spectroscopy. Many iron(III) oxidations of organic species, especially potential ligands with sulphur donor atoms (e.g. thiols, thiomalate), involve such transient complexes, often observable in test-tube experiments and readily characterisable spectroscopically. In cases where the intermediate is more labile, its presence can often be inferred from the kinetics of the reaction. If there is neither spectroscopic nor kinetic evidence for an intermediate, then a one-step mechanism is likely, but by no means proven. It is very difficult to distinguish between an intermediate which is too transient, or present at a concentration too low to detect, and a transition state — a one-step inner-sphere mechanism is just the limiting case. For one-step redox reactions there is still the question of inner-sphere versus outer-sphere electron transfer.

Some idea of the area covered by this section is given by Table 12.5. This gives examples of the types of substrates involved, indicating whether or not intermediate complexes may be involved.

Table 12.5 — Intermediates in aquacation oxidations of simple inorganic and organic ions and molecules

Oxidant	Evidence for intermediate		No evidence for intermediate
	Spectrophotometric	Kinetic	
Tl^{III}	hypophosphite oxalic acid	hypophosphite cycloalkenes	catechol
Mn^{III}	hydrazoic acid hydroquinone	bromide alcohols	hydroxylamine
Fe^{III}	thiosulphate cysteine	sulphite catechol	hydrazine acetoin
Co^{III}	chloride malic acid	hydrogen peroxide α-amino acids	thiocyanate thiourea
Ce^{IV}	hypophosphite acetic acid	bromide glycerol	benzaldehyde lactic acid

12.6 OXOANION OXIDANTS

So far very little has been said about this class of oxidants. Such oxoanions as permanganate, dichromate, and hypochlorite are important oxidants, as are peroxoanions such as peroxodisulphate. Several are particularly powerful (Table 12.6), but often they react slowly even with powerful reductants (see Chapter 8).

Oxoanions can act as one-electron or two-electron oxidants. For oxoanions derived from transition metals, accessible oxidation states may be one unit apart (e.g. MnO_4^{2-}/MnO_4^{-}) and thus one-electron transfer facile. For sp-block elements, stable oxidation states are generally two units apart. Now two-electron transfer will be favoured. These can often be achieved simply by transfer of an oxygen atom, as in, for example:

$$SO_3^{2-} + ClO^- \rightarrow SO_4^{2-} + Cl^-$$

However sulphite can also act as a one-electron reductant:

Table 12.6 — Standard redox potentials for oxoanion and peroxoanion oxidants in aqueous solution ($a_{H^+} = 1.0$; 298.2 K)

Oxoanions	$E^{\ominus}(V)$	Peroxoanions	$E^{\ominus}(V)$
$BrO_4^-,2H^+/BrO_3^-,H_2O$	$+1.76$	$S_2O_8^{2-}/2SO_4^{2-}$	$+2.12$
$ClO^-,2H^+/Cl^-,H_2O$	$+1.70$	$P_2O_8^{4-},2H^+/2HPO_4^{2-}$	$+2.07$
$MnO_4^-,8H^+/Mn^{2+},4H_2O$	$+1.51$	$HSO_5^-,2H^+/HSO_4^-,H_2O$	$+1.84$
$ClO_4^-,8H^+/Cl^-,4H_2O$	$+1.39$		
$Cr_2O_7^{2-},14H^+/Cr^{3+},7H_2O$	$+1.23$		
Cf. $O_2,4H^+/2H_2O$	$+1.23$		

$$SO_3^{2-} - e^- \rightarrow SO_3^{\overline{\cdot}}, \text{ then}$$

$$2\,SO_3^{\overline{\cdot}} \rightarrow S_2O_6^{2-}$$

This path is preferred in reactions with one-electron oxidants such as iron(III) or cerium(IV). One-electron transfer from the NO_2^- or ClO_2^- anions, or to NO_2 or ClO_2, should provide easy redox paths. In these cases, unusually, NO_2 and ClO_2 are stable radicals. Though several such one-electron redox reactions are known, the reaction of NO_2^- with sulphite (as of $ClO + NO_2$ in the gas phase) involves oxygen transfer, i.e. is a two-electron process.

The most frequently studied peroxoanion oxidant is peroxodisulphate, prepared very easily by electrolytic oxidation of potassium sulphate solution:

$$2\,SO_4^{2-} \xrightarrow{-2e^-} S_2O_8^{2-}$$

Despite its very high redox potential (Table 12.6), oxidations by peroxodisulphate are usually slow. Indeed oxidation of water is so slow that aqueous solutions of potassium peroxodisulphate can be kept for days with very little sign of the thermodynamically expected decomposition. The main reason for this inertness is the strong O–O bond which has to be broken *en route* to the normal product, sulphate. There are a variety of mechanisms for oxidation by peroxodisulphate. Which mechanism operates in a given situation depends on such factors as the ease of transfer of an electron from the reductant and the nature and velocity of reactions undergone by the sulphate ion radical intermediates. There are cases where the rate-determining step is simply:

$$S_2O_8^{2-} \rightarrow 2\,SO_4^{\overline{\cdot}}$$

Such reactions are zero-order in the reductant, and in principle all have the same rate constant under given conditions. On the other hand there are several transition metal complexes whose reaction with peroxodisulphate is zero-order in oxidant. Here the mechanism involves rate-determining dissociation of the complex followed by more rapid reaction of the products of dissociation with the peroxodisulphate. Between these extremes there are simple second-order reactions and a range of more complicated kinetic patterns dominated by the kinetic behaviour of chains of radical intermediates. Other peroxoanions, for instance peroxodiphosphate or peroxo-monosulphate, have been much less studied, but probably have similarly complicated patterns of kinetic behaviour.

Both for oxoanions and for peroxoanions, redox reactions can often be greatly accelerated by suitable catalysts. Transition metal aqua-ions are often effective catalysts, acting as efficient intermediaries in the electron transfer process. One well-known example is the role of Mn^{2+} aq in catalysing permanganate oxidations, often manifested in the slowness of reaction in the early stages of titrations using permanganate. Peroxodisulphate oxidations are also susceptible to catalysis by traces of transition metal ions. Indeed it is often difficult to be sure that one has totally avoided such catalytic contributions in kinetic studies of peroxodisulphate oxidations. Fortunately, the addition of a small amount of a strongly complexing ligand such as EDTA sequesters traces of metal ions. The redox properties of the EDTA complexes are sufficiently different (see Chapter 7 for the effects of complex formation on redox potentials) for them not to act as electron-transfer catalysts. In similar vein, complexing agents are added to certain washing powders to prevent traces of metal ions causing premature decomposition of such additives as perborates.

13

Past, present, and future

Since the earliest theories of salts dissociating into ions on dissolution in water and Werner's measurements of conductivities of solutions of his complex salts, the behaviour of simple and complex inorganic ions in solution has been intensively studied. The chemistry involved is still a matter for investigation, discussion, and, in some respects, controversy. A great deal is known, but there is still much to discover and even more to understand.

As indicated in Chapter 1, new aqua-ions are still being discovered and characterised. The range of solvents for studies of ions in solution is also still being extended. In particular, in recent years there has been much interest in the solvent properties of molten salts, including electrolyte mixtures that are liquid at ambient temperatures. Prominent amongst these are the chloroaluminate melts, mixtures of aluminium chloride with a chloride of an organic cation such as pyridinium.

In recent years, NMR spectroscopy and diffraction techniques have yielded a lot of information on the nature and interactions of ions in solution. Advances in NMR instrumentation mean that spectra can be obtained from a much wider range of nuclei, solute and solvent, than previously. From the earliest days of NMR it has been possible to obtain spectra for such nuclei as ^{205}Tl and ^{133}Cs, as well as the ^1H present in almost every solvent. Steady development led to great improvements in NMR of such nuclei as the remaining alkali metals and the ubiquitous solvent nuclei ^{13}C, ^{14}N, ^{15}N, and ^{17}O. In the past few years techniques have been developed for the successful monitoring of even such extremely insensitive nuclei as ^{57}Fe. Alongside these extensions to ever more nuclei, developments in even such exhaustively examined areas as ^1H NMR spectroscopy are leading to further insights into ion solution chemistry. Thus SIMPLE (Secondary Isotope Multiplet NMR of Partially Labelled Entities) distinguishes between bound OH and OH_2 protons in such a way that, with complementary information from ^{17}Al NMR, a full and detailed picture of the solution structure of the complicated polynuclear aluminium(III) species (see section 5.2, especially Fig. 5.3) is now being built up.

Magnetic resonance and diffraction techniques are now being widely applied to the examination of the extent of ion-pairing, the precise geometry of ion-pairs and complex ions in solution, the nature of secondary solvation shells, and structural aspects of the solvent in electrolyte solutions. In Chapters 2 and 10 applications of NMR spectroscopy to the determination of solvation numbers and to the kinetics of solvent exchange were described—both applications restricted to solvated metal ions undergoing slow exchange (on the NMR time-scale). It should be added here that concentration, solvent, and temperature effects on chemical shifts for solvent nuclei in fast-exchange situations can yield much information on ion–solvent and ion–ion interactions. Similar information can also be obtained from relaxation times (spin–lattice and spin–spin, T_1 and T_2 (mainly the former) obtained from line widths of NMR signals. Such results give insight into nuclear–dipole (i.e. in this context ion–solvent) interactions and their perturbation by other ions (i.e. ion-pairing), into reorientation times of solvent molecules and how these are affected by the presence of ions, and into tumbling times for entities such as $[M(OH_2)_x]^{n+}$. Recent examples include ^{29}Si relaxation studies of aqueous alkali metal silicate solutions, where both static (polysilicate structures and polysilicate–cation interactions) and dynamic (species rotation and chemical exchange) information was gleaned, and ^{31}P relaxation times as a probe for interactions between $V(\eta^5-C_5H_5)_2Cl_2$ and nucleotides in solution. It is sometimes possible to estimate internuclear distances from relaxation measurements, e.g. intramolecular H...H distances from spin–lattice relaxation times (T_1). Such approaches, and other NMR, X-ray, and neutron spectroscopic, diffraction, and scattering techniques as yet in their infancy (see Chapter 3), can be expected to improve our understanding of the nature and dynamic behaviour of ions in solution a great deal over the next few years. Some problems may well remain intractable, for example solvation numbers of ions which interact only weakly with solvents. Here rates of exchange may be so fast as to make the application of current models and definitions difficult or impossible.

NMR and diffraction methods are effectively single-ion methods, but the majority of thermodynamic and electrochemical experiments on electrolyte solutions give results which have to be apportioned between the constituent cations and anions. Much debate has led to fairly widely accepted methods for obtaining single-ion values for a range of such parameters in aqueous solution, but the situation with respect to non-aqueous and mixed aqueous media is in many respects much less satisfactory. A general consensus of opinion should soon emerge on such topics as single-ion solvation enthalpies and entropies. The problems of explaining and understanding fully the dependence of such parameters on properties of ions and solvents is likely to exercise many chemists for a long time.

The areas of pK_a values for aqua-metal ions and of stability constants in aqueous solution are well documented by now, though knowledge of analogous quantities in non-aqueous media is still rudimentary. Enormous advances have been made in the computation of equilibrium constants for multivariable systems such as series of polynuclear hydroxo-cations or of ternary complexes. Modelling of bioinorganic or of geochemical systems is now possible to a high degree of complexity. Indeed at the moment the power of the computer appears to have outstripped the precision of instrumental results in some areas. It is often necessary to apply chemical intuition to choose between various sets of computer results that are consistent with the experimental results within their likely uncertainties, though sometimes ancillary results from other experimental approaches can be helpful in choosing the best representation of a given system. Thus, for instance, X-ray and NMR studies of

appropriate solutions can sometimes give valuable clues to major species in equilibria involving hydroxo- or oxo-bridged polynuclear species. Studies of polynuclear
species of this type are now being extended to thio-analogues such as $W_3S_4^{4+}$ aq,
$Mo_3S_4^{4+}$ aq, $Mo_4S_4^{5+}$ aq, $Mo_3O_2S_2^{4+}$ aq, and $Mo_3FeS_4^{4+}$ aq (see end of section 5.2); the
last of these takes us to the frontiers of bioinorganic chemistry.

The study of complexes in solution and their stability constants has been
rejuvenated by the discovery, development, and exploitation of the various families
of macrocyclic ligands described in Chapter 6. The development of ligands ever more
powerful for complexing the alkali metal cations, and ever more selective, is still
vigorous. Thus recent development of spherands (Fig. 13.1(a)) has quickly been
followed by hybrid spherand–cryptand ligands (Fig. 13.1(b)) which show a dramatic
increase in complexing strength and in selectivity over the original crown ether and
cryptand ligands. Bioinorganic chemistry and classical physical inorganic chemistry
have merged in this area. Thus it has now proved possible to use ^{87}Rb NMR
spectroscopy to estimate a stability constant (K_1) for the Rb^+ complex of the
gramicidine transmembrane channel. Kinetic studies are now just being extended to
the transport of alkali metal cations across model membranes, which involves rates
of transport as well as rates of uptake, i.e. of complex formation between cation and
ionophore. The simple ideas of matching ion sizes (radii) to cavity sizes (see Table
6.6) are currently undergoing critical examination and development into more
sophisticated theories.

There have also been parallel advances in the preparation and study of polyaza-
and polythia-macrocyclic ligands, and their complexes with transition metal ions.
Thus the N_4S cryptand shown as Fig. 13.1(c) binds Cu^{2+} strongly and selectively,
while the combined azamacrocycle–catechol ligand shown as Fig. 13.1(d) has been
tailored to be a particularly effective ligand for the Fe^{3+} cation. The tetraaza-
macrocycle or crown ligand shown in Fig. 13.1(e) (R = H or $C_{16}H_{33}$) is selective for
the incorporation of platinum(II); the ligand with R = cetyl, $C_{16}H_{33}$, is particularly
useful as it can be used to extract platinum selectively from aqueous media. Several
dual-purpose ligands of the type shown in Fig. 13.1(f) and 13.1(g) have been
synthesised. These ligands will bind one 'hard' cation and one 'soft' cation. The Fig.
13.1(f) ligand gives, for example, a binuclear $Pt^{II}Cu^{II}$ complex, while the Fig. 13.1(g)
ligand can bind an alkali metal cation or a 'hard' transition metal cation in its cryptate
cavity and an Rh^I–CO entity at the azathia-moiety. Selective complexation of Li^+ by
the organometallic ionophore shown in Fig. 13.1(h) gives a species rather similar to
the alkali cation plus rhodium(I)–carbonyl derivative of the Fig. 13.1(g) ligand.
These and related avenues are being explored with a view to finding reversible
systems where cation uptake properties can be switched on by radiation or electron
transfer. Thus the potential ligand shown in Fig. 13.1(i) (=L) only transports Li^+
after it has undergone one-electron reduction to give the anion radical L . Possibilities of switching by pH change $(L + H^+ \rightleftharpoons LH^+)$ or by oxidative dimerisation
$(2LSH \rightleftharpoons LSSL + 2H^+ + 2e^-)$ are also being explored. Another potentially very
important development is the tailoring of macrocyclic and encapsulating ligands so
that they can be bonded to monoclonal antibodies. The immediate objective is to
facilitate the transport of technetium to tumours for imaging and diagnostic
purposes, the ultimate objective to deliver an appropriate radioisotope for the *in situ*
irradiation and destruction of tumours.

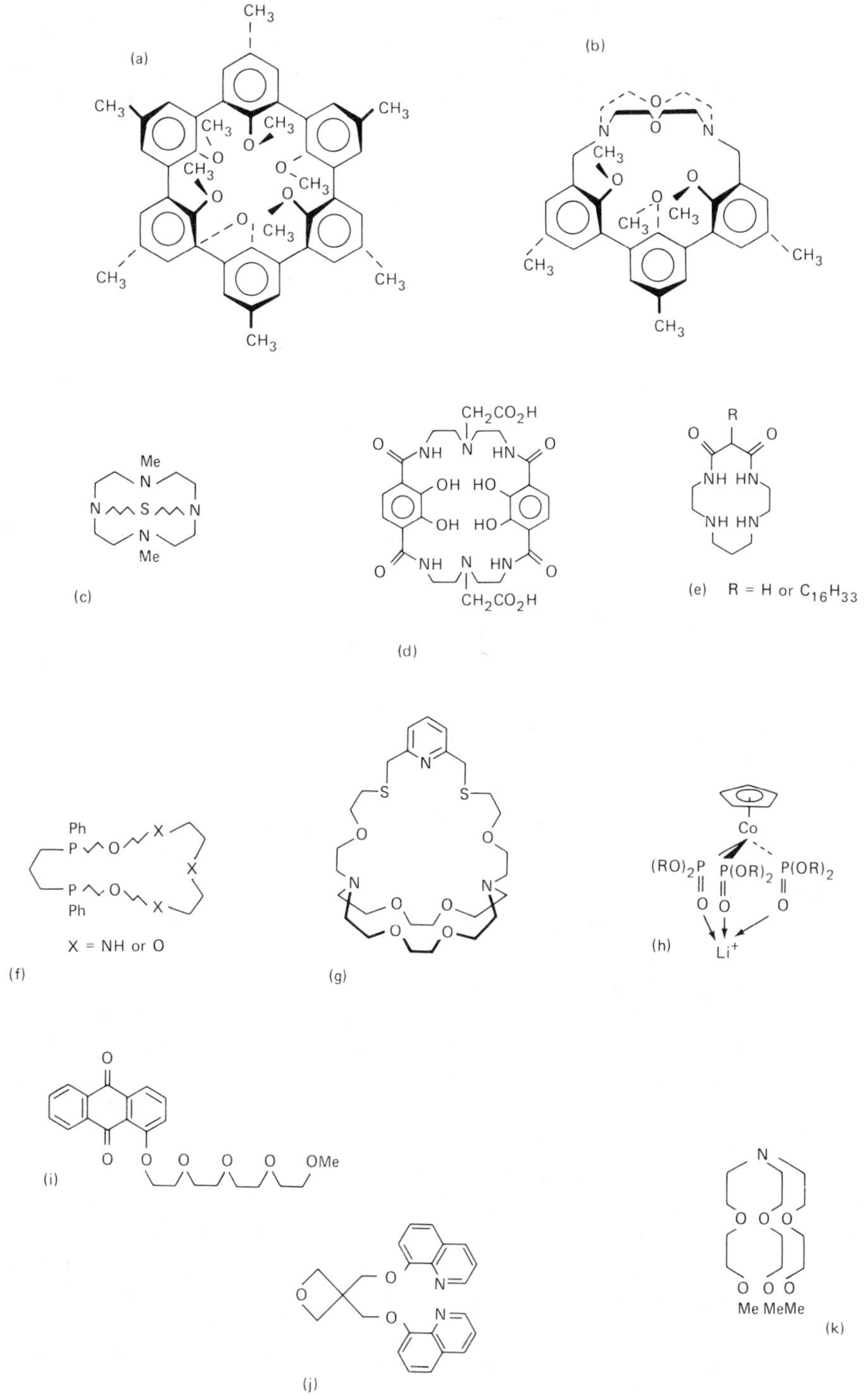

Fig. 13.1 — New macrocyclic and encapsulating ligands.

There are even currently hints that encapsulation, by the pseudo-spherical species C_{60}, may also be a cosmic phenomenon. C_{60} consists of a symmetrical shell of hexagons regularly interspersed with the twelve pentagons demanded by trigonometry (Fig. 13.2; cf. Buckminster Fuller's geodesic domes and a soccer ball). There

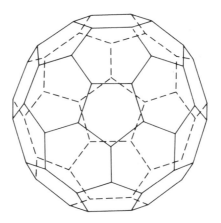

Fig. 13.2 — The species C_{60}.

seems to be some possibility of incorporating a metal ion or atom into its capacious central cavity.

Although the current emphasis in this part of chemistry is on the development of ever more sophisticated macro(poly)cyclic ligands, polydentate non-cyclic analogues have not been entirely neglected. Ligands of the type shown as Fig. 13.1(j) exhibit high selectivity for Li^+ over Na^+ and K^+, while the tripod ligand of Fig. 13.1(k), effective for the specific transport of Ag^+, is being developed industrially for use in phase transfer catalysis. Its complexes are intermediate in stability between those of crown ethers and those of cryptands.

Synthetic advances in the area of polyaza- and polythia-macrocyclic ligands are often linked with efforts to stabilise low, high, or unusual oxidation states of d-block metals (e.g. Cu^I, Cu^{III}, Ni^I, Ni^{III}, Ni^{IV}) in aqueous solution, or to encapsulate a labile metal ion so that redox properties can be examined without worries about ligand dissociation complications. Other multidentate ligands are being developed to enable the synthesis of a comprehensive range of mixed valence complexes for probing fundamental aspects of electron transfer between metal ions, especially in relation to the inner-sphere mechanism for electron transfer. Both in inorganic and in bioinorganic systems, the effects of internuclear separation, of the nature of intervening ligands, and of the medium are only partially established and understood. The Marcus theory of electron transfer, developed for and particularly successful in understanding outer-sphere redox reactions, awaits some extensions to enable it to cope fully with medium effects on electron transfer rates, both inter- and intra-molecularly.

The past decade has seen a great increase in the number of kinetic studies carried out at high pressure (see Fig. 9.2 on page 117). Activation volumes are now available for a variety of reactions involving simple and complex ions in solution. The power of activation volumes in establishing mechanisms is well illustrated by solvent exchange, as discussed in Chapter 9. Applications of high pressure kinetic studies to the investigation of mechanisms of complex formation or redox reactions have been less common, but can be expected to increase in the near future. Activation volumes can be of considerable help in diagnosis of mechanism, but for reactions involving charge separation or diminution the sometimes large consequences of electrostriction have to be borne in mind. However, solvation contributions to measured activation volumes can themselves be of value in probing solvation of transition states when a reaction of well-established mechanism is under scrutiny.

The examples mentioned in the preceding paragraphs by no means cover the whole range of current and expected advances in the study of the multitudinous facets of the chemistry of ionic species in solution. Moreover one can be sure that in years to come all sorts of new facts and insights will emerge from new techniques and theories of which we have no inkling at this time.

Glossary

The reader may well not be familiar with all the scientific and technical terms used in this book, particularly in the first half. Some of the more important of these are therefore explained here. The simplest and most general terms are defined in standard undergraduate textbooks on inorganic or physical chemistry; the more specialised terms are defined in the specialist books cited in the Further Reading section at the end of this book. A few words have acquired specialised meanigns in certain areas of chemistry; their specific meanings in the context of inorganic solution chemistry are given below.

Activation volume. The effect of pressure on rate constants indicates whether the transition state is larger or smaller than the initial state. Activation volumes are obtained from $\Delta V^{\pm} = -RT\{\mathrm{d}(\ln k)/\mathrm{d}P\}$; pressures of up to around 1 kbar are needed in order to get a reasonable change in rate constant. Activation volumes are very useful in mechanism diagnosis when solvation effects can be ignored. This is the case, for example, for solvent exchange at metal ions (see Chapter 9), but often not for redox reactions between charged ions or complexes (see Chapter 12).

Aprotic solvent. A solvent which has no readily ionisable protons. Dimethyl sulphoxide and acetonitrile are classed as aprotic solvents; ethanol is protic only in respect of its hydroxyl proton. Distinction between hydroxylic and non-hydroxylic solvents is sometimes preferable.

Bulk solvent. Strictly this applies to solvent which is unaffected by the presence of solute. However, particularly in relation to NMR studies of ion solvation (Chapter 2), the word 'bulk' is often taken to include anion solvation shells and cation secondary solvation shells, whose signal is generally the weighted mean of all these environments due to fast exchange of solvent molecules between them.

Conjugate acid; *conjugate base*. Conjugate acids and bases are related by the transfer of a proton — RNH_3^+ is the conjugate acid of RNH_2, RNH^- is its conjugate base. Conjugate bases of complex ions can play a key role in mechanisms of base hydrolysis. Thus the rate-determining step in base hydrolysis of cobalt(III)–ammine complexes, e.g. $[CoCl(NH_3)_5]^{2+}$, is believed to be dissociation of the conjugate base, $[CoCl(NH_2)(NH_3)_4]^+$ (see Chapter 11). Although the latter species is present in immeasurably low concentration in aqueous solution, such conjugate bases are detectable in liquid ammonia, and indeed for platinum(IV)–ammine complexes even in aqueous media.

Crystal Field Activation Energy (CFAE). This is the difference between the Crystal Field Stabilisation Energy (CFSE) of the transition state and the CFSE of the initial state for a reaction of a transition metal complex. The former is not easy to estimate, but in general it can be taken that the bigger the CFSE of the reacting complex, the bigger will be the CFAE for its substitution. Thus the particularly high CFSE of low-spin d^6 complexes (cobalt(III); low-spin iron(II)) gives rise to particularly high CFAEs and is an important contributory factor to the very slow rates of substitution at these centres.

Crystal Field Theory. This centrally important theory concerning bonding interactions between metal ions and ligands in transition metal complexes can hardly be compressed into one paragraph here! Readers unfamiliar with CFT, and its close relation Ligand Field Theory, should consult the appropriate chapter in any undergraduate textbook devoted to inorganic chemistry.

Electrostriction. Solvent molecules close to an ion, especially those in a primary solvation shell, will be under the influence of its strong electric field. This generally causes a small but significant decrease in the size of such solvating molecules. Thus the partial molar volume of water (bulk) is $18 \, cm^3 \, mol^{-1}$, but in the primary solvation shell of a 3+ ion is about $15 \, cm^3 \, mol^{-1}$. Electrostriction is important in determining partial molar (molal) volumes of ions in solution. These are negative for ions with large fields (large z^2/r), as the decrease in volume of the water in the hydration shell more than offsets the intrinsic volume of the ion (see Chapter 4.3, Fig. 4.9). Electrostriction is also important in determining activation volumes for reactions involving ions which involve charge cancellation or augmentation, including charge separation in a dissociative rate-determining step.

Hard and Soft Acids and Bases. This classification of metal ions and ligands is discussed in Chapter 6 (see section 6.2).

Hydrodynamic radius. This is the radius of an ion plus such solvent molecules as move through the solution with it. For a given ion in a given solvent it may vary with the method of estimation, depending on the force applied to cause movement.

Hydrolysis. In inorganic aqueous solution chemistry this term is often applied to the reversible loss of one or more protons from an aquacation. Its use is derived from formal equations of this type:

$$[M(OH_2)_6]^{n+} + H_2O \rightleftharpoons [M(OH)(OH_2)_5]^{(n-1)+} + H_3O^+$$

Ionic radii. It would be convenient for the solution chemist to have a list of intrinsic radii of ions, to act as a reference for discussions of hydrodynamic radii, ionic partial molar volumes, and similar quantities. Unfortunately a definitive set of values is not available, and indeed probably cannot exist. In practice there are several sets of ionic radii, for example those of Pauling, of Waddington, of Shannon and Prewitt, or of Goldschmidt out of Wasastjerna. These sets of ionic radii differ significantly from one another. The differences are largest for Li^+, for which radii from 0.60 to 0.94 Å are quoted. The source of the difficulties lies in the fact that until recently ionic radii could only be derived from interatomic distances in crystals, measured by X-ray diffraction. It is easy to draw up self-consistent sets of single-ion radii from such data, but an assumption has to be made before selecting the most acceptable set, which is where the multiplicity of scales mentioned above come from. Recent developments have meant that it has become possible to locate the position of minimum electron density between ions in crystals, thus raising the prospect of *absolute* ionic radii. Unfortunately such radii prove to differ from all the arbitrary scales! Moreover they are not independent of the crystal in which they are measured. It is hardly surprising that the radius of an ion in a range of crystalline compounds varies somewhat with its environment, but it does mean that it is impossible to have an absolute set of ionic radii.

Irving–Williams order. This generalisation about trends in stability constants for metal ions, particularly of the first-row transition metals, is mentioned in Chapter 6.

Jahn–Teller distortion. Orbitally degenerate ground states distort to remove the degeneracy. The ground states concerned for an octahedral complex are t_{2g}^1, t_{2g}^2, e_g^1, etc. The distortions, and the chemical consequences, are greater for e_g states, as in the familiar examples Cr^{2+} ($t_{2g}^3 e_g^1$) and Cu^{2+} ($t_{2g}^6 e_g^3$), than for t_{2g} examples such as Ti^{3+} (t_{2g}^1). However, the Jahn–Teller effect is big enough to affect the visible absorption spectrum of Ti^{3+}aq, as seen in Fig. 4.2. The best explanation for this t_{2g}/e_g difference can be found on pp. 194 and 383 of C. S. G. Phillips and R. J. P. Williams, *Inorganic Chemistry*, Volume 2, Oxford University Press (1966).

Lattice enthalpy. For a concise discussion of lattice enthalpies, lattice energies, and Madelung constants (sufficient for background to section 3.2), see, for example, pp. 285–288 of W. L. Jolly, *Modern Inorganic Chemistry*, McGraw-Hill, 1985.

For a fuller treatment, try D. A. Johnson, *Some Thermodynamic Aspects of Inorganic Chemistry*, 2/e, Cambridge University Press, 1982.

Logarithms. Note the common use of log K values in tabulations of stability constants — these are always decadic logarithms (i.e. logs to base 10).

Madelung constant. The coulombic energy of interaction between an ion and its neighbours in a crystal lattice is $-Az_+z_-e^2/4\pi\varepsilon_0 r$, where r is the distance between ions, ε_0 is the permittivity, e is the charge on the electron, z_+ and z_- are the cation and anion charges, and A is the Madelung constant. A is the sum of a slowly converging series representing the number of neighbours at each of successively increasing interionic distances. The value of A thus depends on the geometry of the crystal.

Marcus–Hush theory. The theory of kinetic–thermodynamic correlation for simple outer-sphere electron transfer reactions (see Fig. 12.8 in Chapter 12).

Oxidative addition. It is often possible to add small molecules to complexes of coordination number four (square-planar) or five and low oxidation state to give an octahedral species with the metal in an oxidation state two units higher. Reactions of alkyl halides with rhodium(I) or iridium(I) complexes provide good examples, for instance

$$trans-[Ir^{I}Cl(CO)(PPh_3)_2] + MeI = Ir^{III}MeClI(CO)(PPh_3)_2$$

Addition of dioxygen, dihydrogen, or fluoroalkenes to iridium(I) or rhodium(I) compounds produces formally octahedral products of indeterminate oxidation state. Oxidative addition reactions are also quite common in the *sp*-block, for example $PBr_3 + Br_2 \rightarrow PBr_5$ or $SnCl_2 + Cl_2 \rightarrow SnCl_4$.

Partial molar (molal) quantities. These play an important role in the description of thermodynamic properties of species in solution. They represent the contribution of a given species to the properties of a system as a whole. Most physical chemistry textbooks introduce this topic through partial molar volumes. When an alcohol is mixed with water, there is a contraction in volume. 20 cm^3 of ethanol plus 80 cm^3 of water give significantly less than 100 cm^3 of mixed solvent. If one adds 1 mole of water to a much greater amount of ethanol, the increase in volume is 14 cm^3. This represents the partial molar volume of the water in the mixture; it is considerably less than the partial molar volume of water itself, which is 18 cm^3 mol^{-1}. In a similar manner, the addition of a salt to water will result in a volume change different from the molar volume of the solid salt. The ions will affect the volume on dissolution through electrostriction of water in their hydration shells. The balance between intrinsic ion sizes, electrostriction, and structural effects in the water may result in positive or negative partial molar volumes for hydrated ions (see Table 4.14). One cannot measure single ion partial molar volumes — they can only be obtained from experimental data (generally densities) by using an extrathermodynamic assumption. The usual assumption is that $\bar{V}(\mathrm{H^+ aq}) = 0$, which is probably not far from the truth (see footnote on page 58). The analogous assumption for ionic partial molar entropies is also probably close to reality, but for other thermodynamic parameters setting the $\mathrm{H^+ aq}$ value at zero may be a very arbitrary choice leading to single ion values far from real-life values. The formal definition of partial molar quantities is $\partial X/\partial n_i$, where n_i is the number of moles of the ith component in the mixture (see any undergraduate textbook of physical chemistry; Barrow gives a reasonably full explanation).

Phase transfer catalysis. Many organic compounds are soluble in a range of organic solvents but not in water, whereas many nucleophiles, such as fluoride or cyanide, are normally available in the form of alkali metal salts which are readily soluble in water but very sparingly soluble in most organic media. For efficient reaction between one species of each category they need to both be in the same solution. Nowadays this is generally arranged by the use of, for instance, a tetraalkylammonium cyanide, or by using a crown ether or cryptand (Chapter 6) to complex the alkali metal cation and render it hydrophobic. This transfer of cyanide, fluoride, or other hydrophilic nucleophile to an organic medium decreases its solvation dramatically, thus increasing its chemical potential and its nucleophilicity and reactivity — hence the term 'phase transfer catalysis'.

pK. $pK = -\log_{10}K$ (analogous to pH). In this book pK values refer specifically to reversible proton loss from an aquacation (so-called 'hydrolysis' — see above).

Polymerisation. In aqueous solution chemistry the word 'polymerisation' is often used rather imprecisely, to cover the formation of oxo- and hydroxo-bridged polynuclear cations. The release of water molecules involved is ignored.

Primary solvation shell. There remains controversy over the precise definition of primary and secondary solvation numbers and shells, especially for those ions where movement of solvent molecules between shells and bulk solvent is very fast. In this book the term 'primary solvation shell' is used to refer to solvent molecules interacting directly with an ion, while 'secondary solvation shell' is used to cover other solvent molecules influenced significantly by the ion.

Protic solvent. A solvent containing ionisable protons (contrast 'aprotic solvent' above).

Radial distribution function. If the probability of finding an entity (in this book an ion or solvent molecule) in a unit small volume is ψ^2, then the probability of finding the entity at a distance between r and $r + \delta r$ from a central reference point will be $4\pi r^2\psi^2\delta r$. This quantity is the product of the probability per unit volume and the volume of space between spheres of radius r and $r + \delta r$. The radial distribution function, or the probability of the entity being at distance r, is thus $4\pi r^2\psi^2$.

Secondary solvation shell. See 'Primary solvation shell' above.

Single-ion properties. It is possible to obtain information on solvation numbers of cations from NMR spectra, on ion–solvent interactions for transition metal cations from ultraviolet–visible absorption spectra, and on hydrodynamic properties of anions or cations individually through moving boundary experiments. But it is impossible to measure single-ion thermodynamic values. Data on electrolyte solutions can be split into self-consistent sets of ionic contributions, but an extrathermodynamic assumption has to be introduced in order to get 'absolute' single-ion values. Such assumptions take several forms, and in many areas are still a matter of some controversy. An early approach was to invoke the similarity in ionic radii (q.v., above) of K^+ and Cl^-, and to equate their single-ion values, in other words to take the value for each as a half that for KCl. The currently most popular form of this assumption is to assume equal values for the large, low-charge, ions Ph_4As^+ (or Ph_4P^+) and Ph_4B^-. An alternative approach frequently encountered is to calculate or estimate a value for one given ion, often H^+, or arbitrarily to set the value for a given ion, e.g. H^+ or R_4N^+, to zero. The choice of single ion assumption can have a dramatic effect on comparisons between values for cations and anions, but whatever the assumption made *differences* between ions of the same charge are always correct.

Stability. This word should be used only in a thermodynamic sense; 'inert' is the correct word for what is loosely called 'kinetically stable' in some quarters. It is advisable to specify what aspect of stability is meant in a particular case — stability to air, to oxidation, to water, to light, to polymerisation, etc. In the case of stability constants for metal complexes (Chapter 6) stability is strictly with respect to dissociation into metal ion plus ligand(s).

Standard states. Thermodynamic quantities with a superscript $\ominus$ refer to processes in which reactants and products are in their standard states. For a pure liquid, e.g. solvent, or solid the standard state is the substance in its stable form at one atmosphere. For a solute, the standard state is a hypothetical ideal solution of unit molality. Standard states are important in relation to the correct form for equilibrium constants, in this book in particular for pK values (Chapter 5) and stability constants for complex formation, especially in connection with the chelate effect (Chapter 6).

Structure of solutions. Water has three-dimensional structure resulting from strong intermolecular hydrogen-bonding. When an organic or inorganic solute is added, the structure of the water may be either enhanced or diminished, depending on the amount as well as the nature of the added species. The classic reference to this is the review by F. Franks and D. J. G. Ives, *Quart. Rev.*, **20**, 1 (1966).

Substitution mechanisms. The basic distinction for nucleophilic substitution, at a
     metal ion in a complex as at carbon, is between dissociative activation and
     associative activation. The incoming group does not interact with the metal or
     the carbon in the former, it does in the latter. These mechanisms were firmly
     established in the first place in organic chemistry (C. K. Ingold, *Structure and
     Mechanism in Organic Chemistry*, Cornell University Press, 1953), where they
     were labelled S_N1 and S_N2 respectively. Many years later, when a fair amount of
     kinetic data had become available on substitution at metal complexes, the
     symbols D, I_d, I_a, and A were introduced in inorganic chemistry (C. H. Langford
     and H. B. Gray, *Ligand Substitution Processes*, Benjamin, 1965). The D and A
     mechanisms, limiting dissociative and associative respectively, are two-stage
     processes with transient intermediates of reduced and increased coordination
     numbers respectively. In between these extremes come the one step processes
     with ordinary transition states. The I_d mechanism corresponds to S_N1, I_a to S_N2;
     the interchange (I) description emphasises that there is no hard and fast
     distinction between associative and dissociative. The relative importance of
     bond making and breaking in transition state formation is continuously variable.
     At the time of writing, IUPAC (International Union of Pure and Applied
     Chemistry) threatens to introduce new nomenclature for reaction mechanisms.
     This will be based on the letters A and D for 'attachment' and 'detachment', with
     subscripts, e.g. N for nucleophilic or nucleofugic. The S_N2/I_a mechanism will
     become A_ND_N, S_N1/I_d will become A_N*D_N, and the D mechanism $A_N + D_N$.
     Increased precision of labelling will inevitably be accompanied by increased
     complexity and cumbersomeness!

Template reaction. This is the preparation of a complex of a complicated ligand from
     a metal ion and two or more ligand precursors. A true template reaction only
     proceeds through the reaction of one precursor with the other actually com-
     plexed to the metal ion.

Term. In the language of spectroscopy, 'term' has the specific meaning of an energy
     level in a system. For a given configuration, for instance the d^n configuration of a
     transition metal ion, there will in general be several 'terms', or discrete energy
     levels. For example a d^2 configuration, as in V^{3+}, has terms ^1S, ^3P, ^1D, ^3F, and
     ^1G. All of these can be involved in electronic transitions, and thus contribute to
     spectra, though selection rules limit the number of transitions between terms
     that are spectroscopically significant. The reader is recommended to consult the
     books by Gerloch and Figgis cited in the Further Reading section at the end of
     this book; G. Herzberg, *Atomic Spectra and Atomic Structure*, Prentice-Hall
     Dover, 1944, gives a very understandable treatment of the atomic term symbols
     from which the transition metal applications develop.

Further reading

The following bibliography gives suggestions for further reading to obtain a fuller account or a different view of various topics. The author's *Metal Ions in Solution* (Ellis Horwood, Chichester, 1978 (paperback 1979)) covers much the same ground as the present book but at a more advanced level. It is fully referenced and therefore a useful source for literature references to specific matters.

GENERAL: SOLUTIONS AND SOLVATION

O. Popovych and R. P. T. Tomkins, *Non-aqueous solution Chemistry*, John Wiley, 1981.
B. E. Conway, *Ionic Hydration in Chemistry and Biophysics*, Elsevier, 1981.
G. W. Nielson and J. E. Enderby (eds), *Water and Aqueous Solutions*, Adam Hilger, 1986.
   These books contain a great deal about many facets of the state and chemistry of ions in solution in various media.

J. N. Murrell and E. A. Boucher, *Properties of Liquids and Solutions*, John Wiley, 1982.
   This book contains a general description of solvents and solvation phenomena.

GENERAL: TECHNIQUES

E. A. V. Ebsworth, D. W. H. Rankin and S. Cradock, *Structural Methods in Inorganic Chemistry*, Blackwell, 1987.
   This deals with most of the techniques mentioned in this book, though not usually in the context of solutions.

W. Kemp, *NMR in Chemistry*, Macmillan, 1986.
   A thorough and readable account of the multitudinous ways in which NMR can help the chemist. The only area missing is ions in solution!

L. Banci, I. Bertini and C. Luchinat, *Magn. Reson. Rev.*, **11**, 1 (1986).
> The suitability of various nuclei for NMR spectroscopy is explained in relation to the relaxation behaviour of the respective nuclei.

R. K. Harris and B. E. Mann (eds.), *NMR and the Periodic Table*, Academic Press, 1978.
> This book gives a good indication of the versatility of NMR as a means of probing ion solvation, especially through the use of NMR signals from various elements scattered throughout the Periodic Table.

AQUA-METAL IONS

Information on these will be found in most undergraduate textbooks on inorganic chemistry.

D. T. Richens and A. G. Sykes, *Comments Inorg. Chem.*, **1**, 141 (1981).
> This article describes the chemistry of the aqua-ions of molybdenum in various oxidation states, and is of particular interest in relation to the topic of recent and forthcoming characterisation of aqua-ions.

ION SOLVATION

J. O'M. Bockris and A. K. N. Reddy, *Modern Electrochemistry*, Plenum-Rosetta, 1973 (2 volumes).
> Chapters 2 and 4 of Volume 1 include general descriptions of ion solvation.

J. E. Desnoyers and C. Jolicoeur, in *Modern Aspects of Electrochemistry*, Volume 5, ed. B. E. Conway and J. O'M. Bockris, Butterworths, 1969, Chapter 1.
J. I. Padova, ibid., Volume 7, 1972, Chapter 1.
P. Kebarle, ibid., Volume 9, 1974, pp. 1–46.
> The first two of these reviews provide a brief overview, the last a more advanced and detailed treatment, of several aspects of the solvation of ions in water and in mixed and non-aqueous solvents.

S. F. Lincoln, *Coord. Chem. Rev.*, **6**, 309 (1971).
A. Fratiello, *Progr. Inorg. Chem.*, **17**, 57 (1972).
> Reviews of the determination of solvation numbers by NMR.

A. I. Popov, *Pure Appl. Chem.*, **41**, 275 (1975).
> A short conference lecture demonstrating how NMR and infrared–Raman spectroscopies give complementary information on ion solvation.

D. E. Irish, in *Ionic Interactions*, Volume 2, ed. S. Petrucci, Academic Press, 1971, Chapter 9.
R. E. Verrall, in *Water—A Comprehensive Treatise*, Volume 3, ed. F. Franks, Plenum, 1973, Chapter 5.
> These chapters detail how vibrational spectroscopy furnishes information on solvated ions.

B. N. Figgis, *Introduction to Ligand Fields,* John Wiley, 1966.
  Chapters 3, 4, 7, and 9 detail ultraviolet–visible spectra of hydrated transition metal cations and explain how ligand field parameters are obtained.

M. Gerloch, *Orbitals, Terms and States,* John Wiley, 1986.
G. Herzberg, *Atomic Spectra and Atomic Structure,* Prentice-Hall/Dover, 1944.
W. G. Richards and P. R. Scott, *Structure and Spectra of Atoms,* John Wiley, 1976, Chapters 3 and 4.
  These books are recommended to readers whose knowledge of the fundaments of electronic spectroscopy is insufficient to cope with the basics used by Figgis.

D. Feakins, R. O'Neill, and E. Waghorne, *Pure Appl. Chem.,* **54,** 2317 (1982).
  A short account of an important recent development in obtaining single ion hydration numbers from ion movement experiments.

SOLVATION NUMBERS

J. F. Hinton and E. S. Amis, *Chem. Rev.,* **71,** 627 (1971).
  Probably the most exhaustive compilation of hydration and solvation numbers. It is, understandably but regrettably, uncritical.

ION–SOLVENT DISTANCES

J. E. Enderby and G. W. Nielson, in *Water—A Comprehensive Treatise,* Volume 6, ed. F. Franks, Plenum, 1979, Chapter 1.
R. Caminiti, G. Licheri, G. Piccaluga and M. Magini, *Rev. Inorg. Chem.,* **1,** 333 (1979).
  Thorough descriptions and reviews of the estimation of ion-solvent distances by X-ray and neutron diffraction studies of concentrated aqueous salt solutions.

H. Ohtaki, *Rev. Inorg. Chem.,* **4,** 103 (1982).
  A detailed report on ion–solvent distances determined by diffraction techniques in solution.

The book by Murrell and Boucher cited in the General: Solutions and Solvation section above includes a useful short account of the application of diffraction techniques to electrolyte solutions. For further information on EXAFS, XANES, and related techniques, see:

B.-K. Teo and D. C. Joy (eds.), *EXAFS Spectroscopy: Techniques and Applications,* Plenum, 1981.
T. K. Sham, *Accts. Chem. Res.,* **19,** 99 (1986).

For links between X-ray diffraction results, thermochemistry of complex formation, and Raman spectra of solvento-ions and complexes, see:

S.-I. Ishiguro and H. Ohtaki, *J. Coord. Chem.,* **15,** 237 (1987).

HYDROLYSIS AND POLYMERISATION

C. F. Baes and R. E. Mesmer, *The Hydrolysis of Cations,* John Wiley, 1976.
    The definitive book on this area—comprehensive and comprehensible.

STABILITY CONSTANTS

H. Rossotti, *The Study of Ionic Equilibria,* Longman, 1978.
F. R. Hartley, C. Burgess, and R. Alcock, *Solution Equilibria,* Ellis Horwood, 1980.
    Both deal with acid–base equilibria as well as with metal ion–ligand equi-
    libria; the latter gives a good idea of how to design experiments and compute
    results in this area.

D. Munro, *Chem. Brit.,* 100 (1977).
    One of the simplest of several articles devoted to the chelate effect and
    misunderstandings and misinterpretations thereof.

S. Ahrland, J. Chatt and N. R. Davies, *Quart. Rev.,* **12**, 265 (1958). R. G. Pearson
    (ed.), *Hard and Soft Acids and Bases,* Dowden, Hutchinson, and Ross,
    Stroudsburg, Pa., 1973.
    The original Class 'a'/'b' classification, and a detailed account of the develop-
    ment of the now ubiquitous Hard and Soft Acids and Bases generalisation.
J. M. Lehn, *Accts. Chem. Res.,* **11**, 49 (1978).
D. J. Cram, *Angew. Chem. Int. Edn.,* **25**, 1039 (1986).
A. M. Sargeson, *Chem. Brit.,* 1979, 23.
    Digestible accounts of crown ethers, cryptates, cage complexes, and other
    macrocyclic ligands and their complexes.

REDOX POTENTIALS

General treatments are given in most inorganic and physical texts. There is a useful
extensive compilation of redox potentials at the end of Volume 2 of C. S. G. Philips
and R. J. P. Williams's *Inorganic Chemistry.* Until recently the most comprehensive
collection of inorganic data was that of G. Milazzo and S. Caroli (*Tables of Standard
Electrode Potentials,* John Wiley, 1978). Just available is another comprehensive
data source edited by A. J. Bard, R. Parsons, and J. Jordan, entitled *Standard
Potentials in Aqueous Solution.* This was commissioned by IUPAC to update W. M.
Latimer's 1942 book (*The Oxidation States of the Elements and their Potentials in
Aqueous Solution,* Prentice-Hall, 1952 (2nd edn)), for decades the source of all
inorganic redox wisdom. The latest IUPAC recommendations on *absolute* electrode
potentials can be found in *Pure Appl. Chem.,* **58**, 956 (1986).

KINETICS AND MECHANISMS: GENERAL

M. L. Tobe, *Inorganic Reaction Mechanisms,* Nelson, 1972.
R. G. Wilkins, *The Study of Kinetics and Mechanism of Reactions of Transition
    Metal Complexes,* Allyn & Bacon, 1974.
J. D. Atwood, *Inorganic and Organometallic Reaction Mechanisms,* Brooks-Cole,
    1985.

D. Katakis and G. Gordon, *Mechanisms of Inorganic Reactions,* John Wiley, 1987.
F. Basolo and R. G. Pearson, *Mechanisms of Inorganic Reactions,* John Wiley, 1967 (2nd edn).

The first four are undergraduate textbooks. Of the two most recent, Katakis and Gordon's book is the more generally recommendable. The strength of Atwood's book lies in its organometallic and related sections rather than in its treatment of mechanisms of reaction of complexes. The classic text by Basolo and Pearson documents all the early fundamental work.

S. F. Lincoln, *Progr. React. Kinet.,* **9**, 1 (1977).

This review gives a good overview of the application of NMR techniques to a range of reaction types.

S. Suvachittanont, *J. Chem. Educ.,* **60**, 150 (1983).
R. van Eldik (ed.), *Inorganic High Pressure Chemistry,* Elsevier, 1986.

The former gives a good brief introduction to the use of activation volumes for the diagnosis of reaction mechanisms. For the full, definitive account of high pressure inorganic kinetics, consult the latter text.

KINETICS AND MECHANISMS: SOLVENT EXCHANGE

A. E. Merbach, *Pure Appl. Chem.,* **21**, 1479 (1982).

KINETICS AND MECHANISMS: COMPLEX FORMATION

This topic is rather sketchily dealt with in the general references listed above. A reasonably full elementary account will be found in Chapter 12 of the present author's *Metal Ions in Solution*; while an understandable introduction and subsequent updates by one of the pioneers in this field are provided in:

R. G. Wilkins, *Accts. Chem. Res.,* **3**, 408 (1970); *Pure Appl. Chem.,* **33**, 583 (1973); *Comments Inorg. Chem.,* **2**, 187 (1983).

The key original reference to the Eigen–Wilkins mechanism is:

R. G. Wilkins and M. Eigen, *Adv. Chem. Ser.,* **49**, 55 (1965).

KINETICS AND MECHANISMS: SUBSTITUTION

At the level adopted in Chapter 11 of the present book, the relevant chapters in the general books recommended above are insufficient.

KINETICS AND MECHANISMS: REDOX

H. Taube, *Electron Transfer Reactions of Complex Ions in Solution,* Academic Press, 1970.
R. D. Cannon, *Electron Transfer Reactions,* Butterworths, 1980.

Index

Chemistry Cassettes

Part of this book is based on the author's *Ions in Solution*, an audio-cassette program published by the Educational Techniques Group Trust of the Royal Society of Chemistry in the **Chemistry Cassettes** series.

Chemistry Cassettes present authoritative accounts of important aspects of chemistry prepared and spoken by distinguished chemists. Designed for both individual and group study they are invaluable for students and teachers.

Each **Chemistry Cassette** is accompanied by a comprehensive workbook containing diagrams, equations, examples and other material discussed by the speaker. In many cases problem sections are included, designed to stimulate thought and test the user's appreciation of the topic.

Chemistry Cassettes are self-pacing and enable users to listen, read and learn at a pace related to their own needs, abilities and understanding.

Details and prices are available from:

The Distribution Centre, Royal Society of Chemistry
Blackhorse Road
Letchworth, Hertfordshire SG6 1HN, UK

Contents

John Burgess has been Lecturer and is currently Senior Lecturer in Chemistry at the University of Leicester, since 1967. He was awarded a B.A. (1960), and an M.A. (1963) in Natural Sciences, from Sidney Sussex College, Cambridge; and a Ph.D. (1963) in Inorganic Kinetics from the University of Cambridge. Dr Burgess is the author of *Metal Ions in Solution* (Ellis Horwood Limited, 1978).

An Historical Perspective

Through the centuries chemists, and alchemists before them, have tabulated
the elements in ways that bring out their interrelations. Alchemists from the
15th to the 18th centuries employed symbols that were partly scientific, partly
mystical, and displayed considerable ingenuity and artistry in numerous
pictures purporting to show chemical and philosophical relationships between
terrestrial materials, heavenly bodies, and their various properties and
attributes. By the time of Diderot's Encyclopaedia, the arrangement
adumbrated the modern Periodic Table. **Lavoisier** developed such systematic
representations, soon followed by **Dalton** (1808), whose Tables of symbols and
atomic weights for the elements have been claimed as "the basis of modern
chemistry". **Newland's** Law of Octaves was quickly followed by **Mendeleev's**
Periodic Table (1872), the progenitor of all modern Periodic Tables. The usual
rectangular array of Periods and Groups is not quite ideal, and several chemists
have suggested arrangements in the form of spirals, concentric circles, and
even an octagonal staircase — all in essence derived from **de Chancourtois's**
"telluric screw" of 1862. The **The Festival of Britain** spiral (1951) form
illustrated here minimises overcrowding at the centre, looks attractive, shows
links between series of elements well, but does not lend itself to the
presentation of atomic numbers, weights, or other data alongside symbols for
the elements. For this one has to return to the rectangular form, as we have
done for the modern version printed on the front endpapers of this book.

Alchemical "Celestial Tree"
Basil Valentine(?), 15th Century

Diderot, 18th Century